Intelligent Control and Learning Systems

Volume 16

Series Editor

Dong Shen, School of Mathematics, Renmin University of China, Beijing, China

The Springer book series Intelligent Control and Learning Systems addresses the emerging advances in intelligent control and learning systems from both mathematical theory and engineering application perspectives. It is a series of monographs and contributed volumes focusing on the in-depth exploration of learning theory in control such as iterative learning, machine learning, deep learning, and others sharing the learning concept, and their corresponding intelligent system frameworks in engineering applications. This series is featured by the comprehensive understanding and practical application of learning mechanisms. This book series involves applications in industrial engineering, control engineering, and material engineering, etc.

The Intelligent Control and Learning System book series promotes the exchange of emerging theory and technology of intelligent control and learning systems between academia and industry. It aims to provide a timely reflection of the advances in intelligent control and learning systems. This book series is distinguished by the combination of the system theory and emerging topics such as machine learning, artificial intelligence, and big data. As a collection, this book series provides valuable resources to a wide audience in academia, the engineering research community, industry and anyone else looking to expand their knowledge in intelligent control and learning systems.

Hongru Ren · Hui Ma · Qi Zhou · Renquan Lu

Intelligent Control of Nonlinear Systems and Nonlinear Multi-Agent Systems with Complex Constraints

Unveiling Cutting-Edge Research and Insights

Springer

Hongru Ren
School of Automation
Guangdong University of Technology
Guangzhou, China

Hui Ma
School of Statistics and Mathematics
Guangdong University of Technology
Guangzhou, China

Qi Zhou
School of Automation
Guangdong University of Technology
Guangzhou, China

Renquan Lu
School of Automation
Guangdong University of Technology
Guangzhou, China

ISSN 2662-5458 ISSN 2662-5466 (electronic)
Intelligent Control and Learning Systems
ISBN 978-981-96-0319-0 ISBN 978-981-96-0320-6 (eBook)
https://doi.org/10.1007/978-981-96-0320-6

This Springer imprint is published by the registered company Springer Nature Singapore Pte Ltd.
The registered company address is: 152 Beach Road, #21-01/04 Gateway East, Singapore 189721, Singapore

This research monograph is humbly dedicated to all my teachers who have kindled the flame of knowledge within me.

Preface

Nonlinear systems are widely used in intelligent manufacturing, electric power, national defense, and other important fields. However, the nonlinear system has the characteristics of complex structure, strong uncertainty, and high coupling, which makes the controller design process very difficult. Therefore, for a long time, nonlinear system control has been the most important and difficult problem in the field of international industrial automation. In recent years, the intelligent control method of nonlinear systems has become the focus of international scholars, and a series of important research results have been obtained. However, at present, the design of intelligent control methods is mostly based on relatively simple environmental assumptions, and less consideration is given to complex constraints such as parts failure and complex external interference. These practical factors greatly limit the application of traditional intelligent control theory in the control of many industrial processes. Therefore, how to give an integrated design method of intelligent adaptive controller for nonlinear systems to overcome the above negative effects has become an urgent scientific problem to be solved.

With the increasing demand of multi-agent system in industrial complex system control and military ground air system collaborative control, the scale and complexity of control system are increasing. Moreover, with the higher requirements for the intelligence of nodes and the sharing of communication links, the multi-agent system in the open network environment is vulnerable to unknown interference, which results in huge security risks to the stable operation of the system. Based on the research of intelligent control of single nonlinear system under complex constraints, in order to solve the above difficult problems, it is urgent to study the intelligent control theory and method for nonlinear multi-agent system under complex constraints.

This monograph provides the latest progress in intelligent control of nonlinear systems under complex constraints, including single nonlinear systems in Part I and nonlinear multi-agent systems in Part II. Chapter 1 of this monograph summarizes the latest development of intelligent control of nonlinear systems and nonlinear multi-agent systems. Then, intelligent controllers are designed for nonlinear systems and nonlinear multi-agent systems with excellent results in the main part. Complex constraints include periodic disturbance, unknown disturbance, input saturation, output constraint, actuator fault, and so on. And control methods include prescribed performance control, event-triggered control, fixed-time control, fault-tolerant control, neural network adaptive control, fuzzy adaptive control, and so on.

Part I: Some developments of intelligent control of nonlinear systems with complex constraints are introduced. In Chap. 2, the adaptive prescribed performance control is discussed for nonlinear systems with actuator failures. In Chap. 3, the approximation-based Nussbaum gain adaptive control is studied for nonlinear systems with periodic disturbances. In Chap. 4, the adaptive dynamic event-triggered control problem is addressed for nonlinear systems using disturbance observer. In Chap. 5, the adaptive fuzzy control issue is investigated for a single-link flexible-joint robotic manipulator with output constraint.

Part II: Some developments of intelligent control of nonlinear MASs with complex constraints are introduced. In Chap. 6, the adaptive fixed-time control is studied for nonlinear MASs with actuator faults. In Chap. 7, the event-triggered adaptive fault-tolerant control method is proposed for MASs with unknown disturbances. In Chap. 8, the adaptive fault-tolerant containment controller is designed for stochastic nonlinear MASs with input saturation. In Chap. 9, the aperiodic sampled-data consensus control approach is studied for nonlinear MASs with constant delay.

Guangzhou, China
July 2023

Hongru Ren
Hui Ma
Qi Zhou
Renquan Lu

Acknowledgements

The authors would like to express their sincere gratitude to Prof. Hongyi Li, Prof. Zhenyou Wang, Prof. Yuanqing Wu, Prof. Hamid Reza Karimi, Dr. Zhijian Cheng, Dr. Guohuai Lin, Mr. Shubo Li, Mr. Changxin Lu, Mr. Junhao Qiu, Mr. Shen Cheng, Mr. Jintao Chen, Mr. Chen Zhang, and Mr. Shitao Duan. We also thank the entire research team at the Guangdong-Hong Kong Joint Laboratory for Intelligent Decision and Cooperative Control, and the Guangdong Provincial Key Laboratory for Intelligent Decision and Cooperative Control for their support and assistance. Special thanks go to the editors and anonymous reviewers for their valuable feedback that significantly improved the quality of this work.

Contents

Chapter 1
Introduction

1.1 Intelligent Control of Nonlinear Systems

Nowadays, the international control academia has formed a relatively complete analysis method and controller design theory system for the research of linear system. However, linear models only idealize and approximate actual objects, and cannot fundamentally reveal nonlinear phenomena. For example, for linear electronic components such as resistors, there may be a phenomenon of "linear distortion" or nonlinear characteristics when affected by factors such as temperature and humidity changes and self-aging. Therefore, nonlinear system is considered to be the most common system form in nature and engineering technology, which has important research significance. The practical systems in important fields such as aerospace, rail transit, ocean exploration, and power system are typical complex nonlinear systems.

Intelligent control is based on control theory, computer science, artificial intelligence, operations research, and other disciplines, and expands relevant theories and technologies. Among them, fuzzy logic, neural network, expert system, genetic algorithm, and other theories are widely used, as well as adaptive control, self-organizing control, and self-learning control technologies. The research object of intelligent control usually has serious uncertainty. The model uncertainty mentioned here contains two meanings: firstly, the model is unknown or poorly understood; the second is that the structure and parameters of the model may vary over a large range. For highly nonlinear control objects, intelligent control methods can often solve the control problems of nonlinear systems.

In recent years, the research on intelligent control of nonlinear systems has become more and more in-depth. For example, the switching threshold event-triggered adaptive control problem was considered for a class of uncertain nonlinear systems (Pang et al. 2022). The adaptive fuzzy tracking control problem was investigated for a class of strict-feedback nonlinear systems with time-varying input delay and full state constraints (Wang et al. 2020). The adaptive fuzzy control problem was addressed for a class of stochastic nonlinear systems with the risk-sensitive performance index

H. Ren et al., *Intelligent Control of Nonlinear Systems and Nonlinear Multi-Agent Systems with Complex Constraints*, Intelligent Control and Learning Systems 16,
https://doi.org/10.1007/978-981-96-0320-6_1

and unknown uncertainties (Ma et al. 2021). The adaptive event-triggered control problem was studied for uncertain nonlinear systems using intermittent output only (Zhang et al. 2022). The problem of adaptive approximation-based tracking control was tackled for a class of unknown high-order nonlinear systems with unknown powers (Liu et al. 2022). The finite-time adaptive fuzzy control problem was addressed for a class of nonstrict-feedback uncertain nonlinear systems via an event-triggered strategy (Wang et al. 2020). The disturbance observer-based adaptive fuzzy control issue was studied for strict-feedback nonlinear systems with finite-time prescribed performance (Qiu et al. 2022). A combined gradient descent Barzilai-Borwein algorithm and radial basis function neural network output tracking control strategy was proposed for a family of nonlinear systems with unknown drift function and control input gain function (Wang et al. 2023). The problem of adaptive fuzzy output tracking control was investigated for a category of high-order nonlinear systems with input saturation, output constraint, and serious uncertainties (Sun et al. 2021). An adaptive output feedback control was studied for nonlinear systems with mismatched uncertainties under input/output quantization (Zhang et al. 2022).

1.2 Intelligent Control of Nonlinear MASs

When intelligent control is extended to MASs, more problems emerge to be solved, and related research is increasingly valued. For example, the problem of adaptive fuzzy decentralized tracking control design was investigated for a class of large-scale interconnected nonlinear MASs with network-induced input time delay (Ma et al. 2021). An adaptive fuzzy decentralized dynamic surface control strategy was studied for the nonstrict-feedback fractional-order nonlinear large-scale MASs with unknown control directions (Zhan et al. 2022). A finite-time adaptive containment control scheme was proposed for a class of uncertain nonlinear MASs subject to mismatched disturbances and actuator failures (Xiao et al. 2022). An event-triggered cluster consensus scheme was presented for heterogeneous nonlinear second-order MASs subject to cyber-attacks, actuator faults, and integral quadratic constraints under directed communication topology containing a directed spanning tree (Guo et al. 2021). For a class of high-order nonlinear MASs with input hysteresis, an adaptive consensus output-feedback quantized control scheme with full state constraints was investigated (Zhu et al. 2022). The problem of leader-following consensus fault-tolerant control was investigated for MASs with time-varying nonaffine nonlinear faults, where the interactive topology is directional (Dong et al. 2021). The problem of fuzzy adaptive consensus tracking control was investigated for nonlinear MASs with unknown nonlinear control gain functions (Li et al. 2022). The adaptive neural network finite-time consensus fault-tolerant control tracking problem was studied for the nonlinear MASs in the nonstrict-feedback form (Dong et al. 2021). The adaptive bipartite containment control problem was investigated for stochastic nonlinear

MASs with an event-triggered mechanism (Zhang et al. 2022). The adaptive neural controller design was studied for a class of uncertain MASs described by ordinary differential equations and beams (Zheng et al. 2021).

1.3 Book Organization

So far, a large number of important and interesting results of intelligent control have been proposed for nonlinear systems and nonlinear MASs. However, there lacks a monograph to provide the up-to-date advances in the literature. Thus, the main purpose of this book is to fill such a gap by providing some recent developments in the intelligent control issues for nonlinear systems and nonlinear MASs. The materials adopted in the book are mainly based on the research results of the authors.

Besides this short Introduction, this book is organized into two main parts; i.e., Part I: Intelligent Control of Nonlinear Systems and Part II: Intelligent Control of Nonlinear MASs.

Some developments on intelligent control of nonlinear systems with complex constraints are presented in Part I, which starts with Chap. 2 and consists of four relevant but independent chapters as follows.

Chapter 2 designs an adaptive neural network output feedback fault-tolerant control algorithm for a class of nonlinear systems with full state constraints and actuator faults. A state observer is constructed to solve the unmeasurable states problem. Unknown nonlinear functions in the systems are approximated by radial basis function neural networks (RBF NNs). By introducing the nonlinear mapping, the systems with state constraints are transformed into novel systems without state constraints. Moreover, a novel performance function is utilized to guarantee that the tracking error converges within a preset time. Meanwhile, the convergence speed can be adjusted through the parameter design. Finally, it is proved that the control algorithm ensures that all signals in the closed-loop systems are semi-globally uniformly ultimately bounded. The effectiveness of the algorithm is verified by a numerical simulation (Qiu et al. 2023).

Chapter 3 considers the Nussbaum gain adaptive control issue for a type of nonlinear systems, in which some sophisticated and challenging problems, such as periodic disturbances, dead zone output, and unknown control direction are addressed. The Fourier series expansion and radial basis function neural network are incorporated into a function approximator to model time-varying disturbed function with a known period in nonlinear systems. To deal with the problems of the dead zone output and unknown control direction, the Nussbaum-type function is recommended in the design of the control algorithm. Applying the Lyapunov stability theory and backstepping technique, the proposed control strategy ensures that the tracking error is pulled back to a small neighborhood of origin and all closed-loop signals are bounded. Finally, simulation results are presented to show the availability and validity of the analysis approach.

Chapter 4 focuses on the adaptive fuzzy dynamic event-triggered control problem for one class of nonlinear systems with input saturation. A dynamic event-triggered mechanism is adopted to save the communication burden of the system, and is integrated into the controller design process based on the backstepping technique. For handling the asymmetric input saturation, an auxiliary system with the same order as the considered system is constructed. Analysis problems caused by unknown continuous functions are efficiently tackled by applying the approximation theory of fuzzy logic systems. It is proven that all the signals of the resulting closed-loop system are bounded via Lyapunov stability analysis. Both numerical and practical simulations are presented to demonstrate the effectiveness of the theoretical results.

Chapter 5 proposes an adaptive fuzzy controller for a single-link flexible-joint robotic manipulator subject to output constraint, in which the fuzzy logic systems are employed to identify the unknown nonlinearity. The barrier Lyapunov function is constructed to guarantee that the output of the single-link flexible-joint robotic manipulator is not to violate its constraint. In order to reduce the occupation of communication channel bandwidth, the event-triggered control law is developed based on a fixed threshold approach. Incorporating the backstepping technique and Lyapunov stability theory, the proposed control strategy ensures all variables in the flexible-joint robotic manipulator are bounded, and the tracking error converges to the origin with a small scale. Simulation results are given to illustrate the efficacy of the developed control strategy.

Some developments on intelligent control of nonlinear MASs with complex constraints are presented in Part II, which starts with Chap. 6 and consists of four relevant but independent chapters as follows.

Chapter 6 considers the adaptive fixed-time consensus problem for a class of nonlinear multi-agent systems (MASs) with actuator faults. To approximate the unknown nonlinear functions in MASs, radial basis function neural networks are used. In addition, the first-order sliding mode differentiator is utilized to solve the "explosion of complexity" problem, and a filter error compensation method is proposed to ensure the convergence of filter error in fixed time. With the help of the Nussbaum function, the actuator failure compensation mechanism is constructed. By designing the adaptive fixed-time controller, all signals in MASs are bounded, and the consensus errors between the leader and all followers converge to a small area of origin. Finally, the effectiveness of the proposed control method is verified by simulation examples (Ren et al. 2023).

Chapter 7 presents an event-triggered consensus control protocol for a class of multi-agent systems with actuator faults, sensor faults, and unknown disturbances. The adaptive neural network compensation control method is introduced to solve the problem of sensor faults. The event-triggered mechanism is developed to reduce the communication burden. In the control design process, the radial basis function neural networks are used to approximate the unknown nonlinear functions, and a nonlinear disturbance observer is used to eliminate the effect of unknown external disturbances. Furthermore, based on the graph theory and Lyapunov stability theory, it is further shown that the consensus tracking errors are semi-globally uniformly ultimately bounded. Finally, the simulation example illustrates the effectiveness of the designed control protocol.

Chapter 8 considers the fault-tolerate containment control problem for stochastic nonlinear multi-agent systems in the presence of input saturation and sensor faults. In order to solve the problem of input saturation, a smooth function is used to approximate the controller saturation function. Due to the excellent approximation characteristic, neural network (NN) is used to deal with unknown nonlinear functions and unknown sensor faults. Meanwhile, by using the auxiliary system and combining adaptive backstepping technique with adaptive NN control, a fault-tolerant containment control approach is proposed. By using graph theory and stochastic Lyapunov stability theory, it can be proved that all signals in the closed-loop system are semi-globally uniformly ultimately bounded in probability. Finally, simulation results are given to show the effectiveness of the proposed control scheme.

Chapter 9 considers the synchronization issue for network systems with nonlinear dynamics. Together with Zero-Order Holder (ZOH), the aperiodic sampled-data control law is utilized. Compared with the traditional periodic sampled-data control method, this approach demonstrates more greater flexibility. Adopting input delay approach, the initial sampled-data system is remodeled by continuous time system involving time-varying delays in the control signals. For the purpose of designing the sampling controllers suffering constant delays, an updated Lyapunov functional is developed from the augmentation of Wirtinger's inequality. Such a Lyapunov functional results in efficient and simplified synchronization conditions. A sufficient condition for synchronizability of network systems is set up. Then, for the case of unstable systems with some constant delays, a fresh discretized Lyapunov functional is introduced. At last, we utilize the numerical simulation outcomes to prove the efficacy and advantage of our algorithm. Moreover, based on the network unmanned ground vehicle systems, the experiment results in a real scenario are provided to illustrate the effectiveness of the designed synchronization scheme.

References

Dong G, Ren H, Yao D, Li H, Lu R (2021) Prescribed performance consensus fuzzy control of multiagent systems with nonaffine nonlinear faults. IEEE Trans Fuzzy Syst 29(12):3936–3946

Dong G, Li H, Ma H, Lu R (2021) Finite-time consensus tracking neural network FTC of multi-agent systems. IEEE Trans Neural Netw Learn Syst 32(2):653–662

Guo XG, Liu PM, Wang JL, Ahn CK (2021) Event-triggered adaptive fault-tolerant pinning control for cluster consensus of heterogeneous nonlinear multi-agent systems under aperiodic DoS attacks. IEEE Trans Netw Sci Eng 8(2):1941–1956

Li Y, Li K, Tong S (2022) An observer-based fuzzy adaptive consensus control method for nonlinear multiagent systems. IEEE Trans Fuzzy Syst 30(11):4667–4678

Liu YH, Liu Y, Liu YF, Su CY, Zhou Q, Lu R (2022) Adaptive approximation-based tracking control for a class of unknown high-order nonlinear systems with unknown powers. IEEE Trans Cybernet 52(6):4559–4573

Ma M, Wang T, Qiu J, Karimi HR (2021) Adaptive fuzzy decentralized tracking control for large-scale interconnected nonlinear networked control systems. IEEE Trans Fuzzy Syst 29(10):3186–3191

Ma M, Sun K, Wang T, Qiu J (2021) Adaptive fuzzy risk-sensitive control for stochastic strict-feedback nonlinear systems with unknown uncertainties. IEEE Trans Fuzzy Syst 29(12):3794–3802

Pang N, Wang X, Wang Z (2022) Event-triggered adaptive control of nonlinear systems with dynamic uncertainties: the switching threshold case. IEEE Trans Circuits Syst II: Express Briefs 69(8):3540–3544

Qiu J, Wang T, Sun K, Rudas IJ, Gao H (2022) Disturbance observer-based adaptive fuzzy control for strict-feedback nonlinear systems with finite-time prescribed performance. IEEE Trans Fuzzy Syst 30(4):1175–1184

Qiu J, Cheng Z, Lin G, Ren H, Lu R (2023) Preseribed performance control for a class of nonlinear pure-feedback systems with actuator faults. J Guangdong Univ Technol 40(2):55–63

Ren H, Ma H, Li H, Wang Z (2023) Adaptive fixed-time control of nonlinear MASs with actuator faults. IEEE/CAA J Automatica Sinica 10(5):1252–1262

Sun W, Su SF, Liu ZG, Sun ZY (2021) Adaptive intelligent control for input and output constrained high-order uncertain nonlinear systems. IEEE Trans Syst Man Cybernet Syst 51(9):5577–5586

Wang A, Liu L, Qiu J, Feng G (2020) Finite-time adaptive fuzzy control for nonstrict-feedback nonlinear systems via an event-triggered strategy. IEEE Trans Fuzzy Syst 28(9):2164–2174

Wang Y, Wang T, Yang X, Yang J (2023) Gradient descent-Barzilai Borwein-based neural network tracking control for nonlinear systems with unknown dynamics. IEEE Trans Neural Netw Learn Syst 34(1):305–315

Wang T, Wu J, Wang Y, Ma M (2020) Adaptive fuzzy tracking control for a class of strict-feedback nonlinear systems with time-varying input delay and full state constraints. IEEE Trans Fuzzy Syst 28(12):3432–3441

Xiao W, Ren H, Zhou Q, Li H, Lu R (2022) Distributed finite-time containment control for nonlinear multiagent systems with mismatched disturbances. IEEE Trans Cybernet 52(7):6939–6948

Zhan Y, Sui S, Tong S (2022) Adaptive fuzzy decentralized dynamic surface control for fractional-order nonlinear large-scale systems. IEEE Trans Fuzzy Syst 30(8):3373–3383

Zhang Z, Wen C, Xing L, Song Y (2022) Adaptive output feedback control of nonlinear systems with mismatched uncertainties under input/output quantization. IEEE Trans Autom Control 67(9):4801–4808

Zhang Y, Sun J, Li H, He W (2022) Event-triggered adaptive bipartite containment control for stochastic multiagent systems. IEEE Trans Syst Man Cybernet Syst 52(9):5843–5852

Zhang Z, Wen C, Xing L, Song Y (2022) Adaptive event-triggered control of uncertain nonlinear systems using intermittent output only. IEEE Trans Autom Control 67(8):4218–4225

Zheng S, Shi P, Wang S, Shi Y (2021) Adaptive neural control for a class of nonlinear multiagent systems. IEEE Trans Neural Netw Learn Syst 32(2):763–776

Zhu G, Li H, Zhang X, Wang C, Su CY, Hu J (2022) Adaptive consensus quantized control for a class of high-order nonlinear multi-agent systems with input hysteresis and full state constraints. IEEE/CAA J Autom Sinica 9(9):1574–1589

Part I
Intelligent Control of Nonlinear Systems

Chapter 2
Prescribed Performance Control with Actuator Failures

Abstract In this chapter, considering a class of nonlinear systems with full state constraints and actuator faults, an adaptive neural network output-feedback fault-tolerant control algorithm with prescribed performance is proposed. A state observer is constructed to solve the unmeasurable states problem. Unknown nonlinear functions in the systems are approximated by radial basis function neural networks (RBF NNs). By introducing the nonlinear mapping, the systems with state constraints are transformed into novel systems without state constraints. Moreover, a novel performance function is utilized to guarantee that the tracking error converges within a preset time. Meanwhile, the convergence speed can be adjusted through the parameter design. Finally, it is proved that the control algorithm ensures that all signals in the closed-loop systems are semi-globally uniformly ultimately bounded. The effectiveness of the algorithm is verified by a numerical simulation.

2.1 Introduction

With the rapid development of control technology, system functions, and components continue to expand. The system structure has become increasingly complex, prompting the urgent need to establish more reliable control methods in the industrial field. However, the linear system control methods of classical control theory cannot cope with the problems of nonlinearity, unknown parameters, and partially unknown system models in the system. Therefore, many researchers have investigated nonlinear systems and proposed control methods such as adaptive control, fuzzy control, and neural network control (Li et al. 2021; Tian et al. 2015; Lin et al. 2022; Ma et al. 2019; Zheng et al. 2020; Ren et al. 2020; Zhou et al. 2019). Among them, adaptive control attracts researchers' attention because of its unique parameter update method. However, adaptive control requires obtaining the mathematical model structure of the controlled object, which makes it difficult to handle nonlinear systems that are hard to model and only have partial structural information. In recent years, with the development of neural network technology, the flexibility and approximation function of neural network for unknown models can well compensate for the limitations of adaptive control. Therefore, the adaptive neural network control method

H. Ren et al., *Intelligent Control of Nonlinear Systems and Nonlinear Multi-Agent Systems with Complex Constraints*, Intelligent Control and Learning Systems 16,
https://doi.org/10.1007/978-981-96-0320-6_2

that combines neural network with adaptive control has received wide attention (Ma et al. 2020; Zeng et al. 2000; Zhou et al. 2019).

For most adaptive neural network control methods, their control effects largely depend on the controller design and the selection of various parameters. In order to make the system meet the desired performance requirements and achieve smaller steady-state errors and higher convergence rates, researchers have proposed various performance-constrained control methods, such as barrier Lyapunov functions, funnel control, and specified performance control (Li and Tong 2018; Ren et al. 2010; Zhang and Yang 2017; Wang and Yang 2018; Kostarigka and Rovithakis 2012; Ni et al. 2019; Yang et al. 2019; Zhao et al. 2021). For a class of non-strict feedback interconnected nonlinear systems, Li and Tong (2018) proposed an adaptive neural network decentralized control method with specified performance. For multi-agent systems, Yang et al. (2019) designed a specified performance function to constrain the tracking errors of each subsystem, thus achieving the expected tracking effect. Based on the quadrotor unmanned aerial vehicle dynamic model, Zhao et al. (2021) proposed an adaptive specified performance control algorithm. In addition, due to problems such as system states being hard to observe directly or costly to measure, it is very important to study state-unobservable systems (Wang et al. 2016, 2018; Zhou et al. 2013; Yang et al. 2016; Arefi et al. 2014). For instance, Wang et al. (2018) studied the finite-time tracking problem of a class of nonlinear quantized systems with state unobservability. Wang et al. (2016) proposed a new state observer to solve the output-feedback control problem of a class of pure feedback nonlinear systems. Although researchers have obtained many research results in the adaptive neural network specified performance output-feedback control of nonlinear systems, there is still a lack of relevant analysis and research for the situation where actuator faults and full state constraints occur simultaneously in practice.

Due to the combined influence of factors such as component wear, temperature and humidity changes, external disturbances, etc., the probability of actuator faults in real systems is increasing. Actuator faults not only degrade the control performance but also make it challenging to ensure the system stability. Therefore, the research on the control methods for systems with actuator faults has important practical significance. Mao et al. (2016) proposed an adaptive output-feedback control method for a class of multi-input single-output actuator fault systems with unknown control directions. Li and Tong (2017) proposed a novel adaptive neural network output-feedback fault-tolerant control method for a class of uncertain nonlinear interconnected large systems with actuator faults and unknown states, by combining the adaptive backstepping design principle with the Nussbaum gain function characteristics. In addition, in order to ensure the safe operation of the system, the system states need to satisfy certain constraint conditions. Li et al. (2019) proposed an adaptive neural network control algorithm based on barrier Lyapunov function for a class of nonlinear systems with full state constraints, which makes the system states meet the constraint requirements and improves the system stability. Based on the analysis of the above references, it is important to consider actuator faults and full state constraints in the adaptive neural network specified performance control design. Therefore, this chapter proposes a new adaptive neural network output-feedback specified performance control method for a class of pure feedback nonlinear systems with actuator faults and full state constraints.

The main contributions of this chapter are summarized as follows:

1. Compared with the specified performance control design in Yang et al. (2019), Zhao et al. (2021), this chapter can control the convergence time and rate by designing the parameters in the performance function, thus having higher flexibility.
2. Compared with Yang et al. (2019), Li et al. (2019), this chapter uses a nonlinear mapping method to transform the constrained system into a new unconstrained system, thus avoiding the need to make bounded assumptions on the virtual controller, reducing the system conservatism.
3. This chapter considers a class of nonlinear pure feedback systems under multiple constraint conditions of unknown states, specified performance, actuator faults, and full state constraints, and solves their control problems. In addition, this chapter uses dynamic surface technique to avoid the "explosion of computation" problem in backstepping method, making the control design process simpler.

2.2 Preliminaries and Problem Description

2.2.1 System Description

In this chapter, the following n-order pure-feedback nonlinear systems are considered:

$$\begin{cases} \dot{x}_m = f_m\left(\overline{x}_{m+1}\right) + x_{m+1}, \ 1 \leqslant m \leqslant n-1 \\ \dot{x}_n = f_n(x) + u_{\mathrm{f}} \\ y = x_1, \end{cases} \tag{2.1}$$

where $\overline{x}_{m+1} = \left[x_1, x_2, \ldots, x_{m+1}\right]^{\mathrm{T}} \in \mathbf{R}^{m+1}$ and $x = [x_1, x_2, \ldots, x_n]^{\mathrm{T}}$ are the state vectors of the system; $y \in \mathbf{R}$ and $u_{\mathrm{f}} \in \mathbf{R}$ are the system output and input, respectively. $f_m\left(\overline{x}_{m+1}\right)$ and $f_n(x)$ are unknown smooth nonlinear functions. All system states $x_m (m = 1, \ldots, n)$ need to be guaranteed in an open set $\Omega_{x_m} = \left\{x_m : -\underline{k}_{\mathrm{cm}} < x_m < \bar{k}_{\mathrm{cm}}\right\}$, where $\underline{k}_{\mathrm{cm}}$ and $\bar{k}_{\mathrm{cm}}$ are known state boundaries of the system.

Combining the possible bias and gain faults of the actuator, the actuator fault model of the system considered in this chapter is described as

$$u_{\mathrm{f}}(t) = b(1-s)u(t) + \varpi(t),$$

where $u_{\mathrm{f}}(t)$ is the actual controller output; $u(t)$ is the controller input; b is the unknown control direction. $\varpi(t)$ is the additive fault of the controller, and it is a bounded unknown nonlinear function. s represents the unknown fault index of the controller, $0 < s < 1$.

2.2.2 Radial Basis Function Neural Network

In this chapter, radial basis function neural networks (RBF NNs) are used to approximate the nonlinear functions in the system (Yang et al. 2019)

$$f(Z) = \widehat{W}^{\mathrm{T}} S(Z),$$

where $S(Z) = [S_1(Z), S_2(Z), \ldots, S_k(Z)]^{\mathrm{T}}$ is the basis function vector, k is the number of nodes of RBFNNs, $\widehat{W} = \left[\widehat{W}_1, \widehat{W}_2, \ldots, \widehat{W}_k\right]^{\mathrm{T}} \in \mathbf{R}^k$ is the weight vector. There exists a constant ideal weight vector $W = \arg\min_{W \in \mathbf{R}^k} \left\{\sup_{Z \in \Omega} \left| f(Z) - \widehat{W}^{\mathrm{T}} S(Z) \right|\right\}$, such that the following equation holds:

$$f(Z) = W^{\mathrm{T}} S(Z) + \delta(Z), \quad \forall Z \in \Omega \subset \mathbf{R}^q,$$

where $W = [W_1, W_2, \ldots, W_k]^{\mathrm{T}} \in \mathbf{R}^k$. The approximation error $\delta(Z)$ satisfies $|\delta(Z)| \leqslant \varepsilon$, $\varepsilon > 0$. The Gaussian basis function used in this chapter is

$$S_i(Z) = \exp\left[-\frac{(Z - \iota_i)^{\mathrm{T}} (Z - \iota_i)}{\zeta_i^2}\right],$$

where $\iota_i = \left[\iota_{i1}, \iota_{i2}, \ldots, \iota_{iq}\right]^{\mathrm{T}}$ and $\zeta_i (i = 1, 2, \ldots, k)$ are the center and width of the Gaussian function, respectively.

Lemma 2.1 (Yang et al. 2019) *Let* $S\left(\overline{x}_q\right) = \left[S_1\left(\overline{x}_q\right), S_2\left(\overline{x}_q\right), \ldots, S_k\left(\overline{x}_q\right)\right]^{\mathrm{T}}$ *be the basis function vector of RBFNNs, where* $\overline{x}_q = \left[x_1, x_2, \ldots, x_q\right]^{\mathrm{T}}$. *There exists any positive integer* $p \leqslant q$, *such that*

$$\left\| S\left(\overline{x}_q\right)\right\|^2 \leqslant \left\| S\left(\overline{x}_p\right)\right\|^2.$$

2.2.3 Unknown Gain Handling

This chapter introduces Nussbaum function to deal with unknown gain problem, and the related definition and lemma are as follows.

Definition 2.1 (Boulkroune 2016) If a continuous function $N(\xi) : R \to R$ has the following properties:

$$\lim_{s\to\infty}\sup \frac{1}{s}\int_0^s N(\xi)\mathrm{d}\xi = +\infty$$

$$\lim_{s\to\infty}\inf \frac{1}{s}\int_0^s N(\xi)\mathrm{d}\xi = -\infty$$

it is called a Nussbaum function.

Lemma 2.2 (Boulkroune 2016) *Let $V(\cdot)$ and $\xi(\cdot)$ be smooth functions defined on $[0, t_\mathrm{f})$, satisfying $\forall t \in [0, t_\mathrm{f})$, $V(t) \geqslant 0$. It is known that $N(\xi)$ is a smooth Nussbaum-type function. If the following inequality holds, then $V(t), \xi(t)$ and $\int_0^t g(x(\tau))N(\xi)\dot{\xi}\mathrm{d}\tau$ are bounded on the interval $[0, t_\mathrm{f})$, i.e.*

$$0 < V(t) \leqslant c_0 + \mathrm{e}^{-c_1 t}\int_0^t (g(x(\tau))N(\xi) + 1)\dot{\xi}\mathrm{e}^{c_1 t}\,\mathrm{d}\tau,$$

where c_1 is a positive number, c_0 is an appropriate constant. $g(x(\tau))$ is a bounded time-varying function, and takes values in a closed unknown interval $I := \left[l^-, l^+\right] \not\ni 0$.

The control objectives of this chapter are as follows: (1) To ensure that the output $y(t)$ can track the desired target function $y_\mathrm{d}(t)$, $|y_\mathrm{d}(t)|$ bounded by Y_0, the tracking error $\tilde{x}_1 = y - y_\mathrm{d}$ needs to satisfy specified performance; (2) To ensure that the system states are always within the constraint range; (3) To ensure that all signals in the system are semi-global uniformly ultimately bounded.

2.3 Adaptive Fault-Tolerant Control Algorithm Design

In order to facilitate the design of state observer, the following transformation is performed on system (2.1): define $\chi = \frac{x}{\beta} = [x_1/\beta, x_2/\beta, \ldots, x_n/\beta]^\mathrm{T} = [\chi_1, \chi_2, \ldots, \chi_n]^\mathrm{T}$, $\bar{\chi}_{m+1} = \left[\chi_1, \chi_2, \ldots, \chi_{m+1}\right]^\mathrm{T}$, where $\beta = b(1-s)$. Then system (2.1) is equivalent to the following system:

$$\begin{cases} \dot{\chi}_m = \frac{f_m(\bar{\chi}_{m+1})}{\beta} + \chi_{m+1} \\ \dot{\chi}_n = \frac{f_n(\chi)}{\beta} + u + \frac{\varpi}{\beta} \\ \dot{y} = \beta x_2 + f_1(\bar{x}_2), \end{cases} \tag{2.2}$$

where $m = 1, \ldots, n-1$. After the above transformation, x_m is converted to x_m/β, which aims to separate the actuator fault part in the original controller. For system (2.2), after using variable transformation, the coefficient of actual control input u becomes 1, which enables the subsequent design of state observer.

2.3.1 State Observer Design

For system (2.1), considering the state unmeasurable problem, the following state observer is established to estimate the state:

$$\begin{cases} \dot{\hat{\chi}}_m = \hat{\chi}_{m+1} + l_m e_1, & 1 \leqslant m \leqslant n-1 \\ \dot{\hat{\chi}}_n = u + l_n e_1, \end{cases} \tag{2.3}$$

where l_m satisfy the Hurwitz polynomial $p(s) = s^n + l_1 s^{n-1} + \cdots + l_n$. Define the observation error $e = \chi - \hat{\chi}$, $\hat{\chi} = [\hat{\chi}_1, \hat{\chi}_2, \ldots, \hat{\chi}_n]^{\mathrm{T}}$, $e = [e_1, e_2, \ldots, e_n]^{\mathrm{T}}$, and according to Eqs. (2.2) and (2.3), obtain the following state observation error form:

$$\dot{e} = Ae + \frac{F(\chi)}{\beta} + B\frac{\varpi}{\beta}, \tag{2.4}$$

where $A = \begin{bmatrix} -\bar{l}_i & I_{n-1} \\ -l_n & 0_{1\times n-1} \end{bmatrix}$, $\bar{l}_i = [l_1, l_2, \ldots, l_{n-1}]^{\mathrm{T}}$. In addition, there exists $PA + A^{\mathrm{T}}P = -\upsilon I$, $P = P^{\mathrm{T}} > 0$, $\upsilon > 0$ as a design constant. $B = [\underbrace{0, \ldots, 0}_{n-1}, 1]^{\mathrm{T}}$, $F(\chi) = [f_1(\overline{\chi}_2), f_2(\overline{\chi}_3), \ldots, f_n(\chi)]^{\mathrm{T}}$.

Since $f_m(\overline{x}_{m+1})$ is an unknown nonlinear function, it cannot be directly used for controller design. Therefore, using RBF NNs to approximate $f_m(\overline{x}_{m+1})$, we have

$$f_m = W_{0m}^{\mathrm{T}} S_0(Z) + \delta_{0m}(Z), \quad \delta_{0m}(Z) \leqslant \bar{\varepsilon}_{0m}, \tag{2.5}$$

where $Z = x$, $W_{0m} \in R^n$, $S_0(Z) = S_0 = [S_{01}, S_{02}, \ldots, S_{0n}]^{\mathrm{T}}$, $\delta_{0m}(Z)$ is the approximation error, $\bar{\varepsilon}_{0m}$ is the approximation error upper bound of f_m . In addition, there exists a compact set Ω_x, which contains the range of all state x trajectories. According to Eq. (2.5), there exists

$$F(\chi) = W_0^{\mathrm{T}} S_0(Z) + \delta_0(Z), \|\delta_0(Z)\| \leqslant \bar{\varepsilon}_0,$$

where $\bar{\varepsilon}_0 = \sqrt{\bar{\varepsilon}_{01} + \bar{\varepsilon}_{02} + \cdots + \bar{\varepsilon}_{0n}}$, $W_0 = [W_{01}, W_{02}, \ldots, W_{0n}]$, $\delta_0(Z) = \delta_0 = [\delta_{01}(Z), \delta_{02}(Z), \ldots, \delta_{0n}(Z)]^{\mathrm{T}}$. Therefore, combining Eq. (2.4), we obtain

$$\dot{e} = Ae + \frac{1}{\beta}\left(W_0^{\mathrm{T}} S_0 + \delta_0\right) + B\frac{\varpi}{\beta}. \tag{2.6}$$

Choose the following Lyapunov function

$$V_e = e^{\mathrm{T}} P e. \tag{2.7}$$

Combining Eq. (2.6), we obtain

$$\dot{V}_e = -ve^{\mathrm{T}}e + \frac{2}{\beta}e^{\mathrm{T}}P\left(W_0^{\mathrm{T}}S_0 + \delta_0\right) + 2e^{\mathrm{T}}PB\frac{\varpi}{\beta}. \tag{2.8}$$

Define the adaptive parameter as

$$\theta^* = \max\left\{n\left\|W_0\right\|^2, \left\|W_m\right\|^2, \quad 1 \leqslant m \leqslant n-1\right\},$$

where W_m is the ideal weight vector of RNF NNs, define $\hat{\theta}$ as the estimate of θ^*, and $\tilde{\theta} = \theta^* - \hat{\theta}$. According to the definition of the Gaussian basis function of RBF NNS, we have $S_{0i} = \exp\left[-\left[\left((Z-\iota_{0i})^{\mathrm{T}}(Z-\iota_{0i})\right)/\zeta_0^2\right]\right], (i = 1, 2, \ldots, n)$, where $\iota_{0i} = \left[\iota_{0i1}, \iota_{0i2}, \ldots, \iota_{0iq}\right]^{\mathrm{T}}$ is the center of the Gaussian function, ω_0 is the width of the Gaussian function. Therefore, $0 < S_{0i} \leqslant 1$, $S_0^{\mathrm{T}}S_0 = \sum_{i=1}^{n} S_{0i}^2 \leqslant n$. Combining the definition of θ^* and Eq. (2.8), we obtain

$$\frac{2}{\beta}e^{\mathrm{T}}P\left(W_0^{\mathrm{T}}S_0 + \delta_0\right) \leqslant 2\eta e^{\mathrm{T}}e + \eta^{-1}\|P\|^2\theta^* + \eta^{-1}\bar{\varepsilon}_0^2 \tag{2.9}$$

$$\frac{2}{\beta}e^{\mathrm{T}}PB\varpi \leqslant \frac{1}{\varrho}e^{\mathrm{T}}e + \varrho\|P\|^2\bar{\varpi}^2, \tag{2.10}$$

where η, $\bar{\varpi}$ and ϱ are positive constants, and $|\varpi(t)| < \bar{\varpi}$. Substituting Eqs. (2.9) and (2.10) into Eq. (2.8), we obtain

$$\dot{V}_e \leqslant -\left(v - 2\eta - \frac{1}{\varrho}\right)e^{\mathrm{T}}e + \eta^{-1}\|P\|^2\theta^* + \varrho\|P\|^2\bar{\varpi}^2 + \eta^{-1}\bar{\varepsilon}_0^2. \tag{2.11}$$

2.3.2 Adaptive Controller Design

This section will use nonlinear mapping for full state constraints, and design the controller and adaptive law of nonlinear system with full state constraints and specified performance by using specified performance design to make tracking error meet expected performance requirements.

Following Zhang et al. (2017), for the full state constraint problem of system (2.1), we perform the following coordinate transformation on the state observer (2.3):

$$\begin{cases} s_2 = \ln\frac{\underline{K}_2+\hat{\chi}_2}{\bar{K}_2-\hat{\chi}_2} \\ \vdots \\ s_n = \ln\frac{\underline{K}_n+\hat{\chi}_n}{\bar{K}_n-\hat{\chi}_n}, \end{cases} \tag{2.12}$$

where $\underline{K}_m$ is the lower bound of $\hat{\chi}_m$ constraint, $\bar{K}_m$ is the upper bound of $\hat{\chi}_m$ constraint, $m = 1, 2, \ldots, n$.

Remark 2.1 This method transforms the constrained system into an unconstrained system by coordinate transformation, and then designs the controller on the new system. Compared with the full state constraint in literature (Yang et al. 2019; Li et al. 2019) using barrier Lyapunov function, it not only does not need to assume that the virtual controller is bounded, reducing conservatism, but also simplifies the process, reducing the difficulty of designing state constraint control for the system.

The specified performance is described as follows:

$$-\mu(t) < x_1(t) < \mu(t), \quad \forall t > 0,$$

where the performance function $\mu(t)$ is a bounded and monotonically decreasing smooth positive function. We select the following function as the performance function.

$$\mu(t) = \begin{cases} \ln\left(C_0\left(t_f - t\right)^{\bar{n}} + 1\right) + \mu_\infty, & 0 < t < t_\mathrm{f} \\ \mu_\infty, & t \geqslant t_\mathrm{f}, \end{cases}$$

where $C_0 = \frac{e^{\varsigma} - 1}{\left(t_f - t_0\right)^{\bar{n}}}$. ς, $\bar{n}$ and μ_∞ are all positive design parameters.

Remark 2.2 Compared with literature (Kostarigka and Rovithakis 2012; Ni et al. 2019), the performance function used in this chapter can set the parameter t_f, so that the constraint error can converge within $(-\mu_\infty, \mu_\infty)$ range within time t_f. It can also change the convergence speed of error by setting parameter $\bar{n}$.

We select the following error transformation function according to Liu et al. (2021):

$$\xi_1(t) = \frac{\tilde{x}_1}{\sqrt{\mu^2 - \tilde{x}_1^2}} \tag{2.13}$$

Remark 2.3 Most specified performance designs make $|\tilde{x}_1| < F_\mu$, where F_μ is the performance function. This error transformation function can make $|\tilde{x}_1| \leqslant \sqrt{\frac{\phi^2}{1+\phi^2}} F_\mu < F_\mu$. This shows that under the same F_μ condition, this method can improve the error constraint ability. ϕ is related to the design of control parameters, which will be proved later.

We differentiate $\xi_1(t)$ to get

$$\dot{\xi}_1(t) = \Gamma_1\left(\dot{x}_1 - \frac{\tilde{x}_1\dot{\mu}}{\mu}\right) = \Gamma_1\left(\beta\left(s_2 + e_2\right) + F_1 - \dot{y}_d - \frac{\tilde{x}_1\dot{\mu}}{\mu}\right), \tag{2.14}$$

where $\Gamma_1 = \frac{\mu^2}{\sqrt{\left(\mu^2 - \tilde{x}_1^2\right)^3}}$, $F_1 = \beta\left(\bar{K}_2 - \frac{\underline{K}_2 + \bar{K}_2}{e^{s_2+1}} - s_2\right) + f_1\left(\overline{x}_2\right).$

We define the following coordinate transformation:

$$\begin{aligned} z_m &= s_m - \omega_m \\ \lambda_m &= \omega_m - \alpha_{m-1}, \quad 2 \leqslant m \leqslant n, \end{aligned} \tag{2.15}$$

where z_m is the virtual error surface, ω_m is the filter output, α_{m-1} is the virtual control signal. Based on dynamic surface technology, we define a first-order filter $\tau_m \dot{\omega}_m + \omega_m = \alpha_{m-1}, \omega_m(0) = \alpha_{m-1}(0)$, where τ_m is a positive design parameter, λ_m is the filter error. From $\lambda_m = \omega_m - \alpha_{m-1}$, we can get $\dot{\omega}_m = -\frac{\lambda_m}{\tau_m}$, then $\dot{\lambda}_m = -\frac{\lambda_m}{\tau_m} + M_m(\cdot), M_m(\cdot) = -\dot{\alpha}_{m-1}$ is a continuous function.

Combining formula (2.3), formula (2.12), formula (2.14) and formula (2.15), we can get

$$\begin{aligned} \dot{\xi}_1 &= \Gamma_1 \left(\beta (s_2 + e_2) + F_1 - \dot{y}_d - \frac{\tilde{x}_1 \dot{\mu}}{\mu} \right) \\ \dot{z}_m &= Q_m (G_m + s_{m+1} + l_m e_1) - \dot{\omega}_m \\ \dot{z}_n &= Q_n (u + l_n e_1) - \dot{\omega}_n, \end{aligned} \tag{2.16}$$

where $Q_m = \frac{e^{s_m} + e^{-s_m} + 2}{\underline{K}_m + \bar{K}_m}$, $G_m = \bar{K}_{m+1} - \frac{\underline{K}_{m+1} + \bar{K}_{m+1}}{e^{s_{m+1}} + 1} - s_{m+1}, m = 2, \ldots, n-1$. $Q_n = \frac{e^{s_n} + e^{-s_n} + 2}{\underline{K}_n + \bar{K}_n}$.

The following are the design steps of adaptive fault-tolerant control algorithm:

Step 1: Choose the Lyapunov function for the first step

$$V_1 = \frac{1}{2}\xi_1^2 + \frac{1}{2r_0}\tilde{\theta}^2, \tag{2.17}$$

where r_0 is a positive design parameter, and from Eqs. (2.16) and (2.17) we have

$$\begin{aligned} \dot{V}_1 = \xi_1 \dot{\xi}_1 - \frac{1}{r_0}\tilde{\theta}\dot{\hat{\theta}} = \xi_1 \Gamma_1 \Big(& \beta (z_2 + \lambda_2 + \alpha_1 + e_2) + \\ & F_1 - \dot{y}_d - \frac{\tilde{x}_1 \dot{\mu}}{\mu} \Big) - \frac{1}{r_0}\tilde{\theta}\dot{\hat{\theta}}. \end{aligned} \tag{2.18}$$

Using RBF NNs to approximate $F_1, F_1 = W_1^{\mathrm{T}} S_1(Z_1) + \delta(Z_1)$, where $Z_1 = [x_1, x_2, s_2]^{\mathrm{T}}$, $\delta(Z_1)$ is the approximation error and $|\delta(Z_1)| \leqslant \bar{\varepsilon}_1, \bar{\varepsilon}_1$ is the upper bound of the approximation error of F_1. Using Young's inequality and Lemma 2.1, we obtain

$$\begin{aligned}\xi_1\Gamma_1 F_1 =& \xi_1\Gamma_1\left(W_1^{\mathrm{T}} S_1(Z_1)+\delta(Z_1)\right)\\ \leqslant& \frac{\xi_1^2\Gamma_1^2\theta^* S_1^{\mathrm{T}}(Z_1) S_1(Z_1)}{2a_1^2}+\frac{a_1^2}{2}+\frac{\xi_1^2\Gamma_1^2}{2}+\frac{\bar{\varepsilon}_1^2}{2}\\ \leqslant& \frac{\xi_1^2\Gamma_1^2\theta^* S_1^{\mathrm{T}}(x_1) S_1(x_1)}{2a_1^2}+\frac{a_1^2}{2}+\frac{\xi_1^2\Gamma_1^2}{2}+\frac{\bar{\varepsilon}_1^2}{2}\end{aligned} \tag{2.19}$$

$$\xi_1\Gamma_1\beta(z_2+\lambda_2)\leqslant\xi_1^2\Gamma_1^2\bar{\beta}^2+\frac{\lambda_2^2}{2}+\frac{z_2^2}{2} \tag{2.20}$$

$$\xi_1\Gamma_1\beta e_2\leqslant\frac{\xi_1^2\Gamma_1^2\bar{\beta}^2}{2}+\frac{1}{2}\|e\|^2, \tag{2.21}$$

where a_1 is a positive design parameter, $\bar{\beta}$ is the upper bound of β. Since $\beta=b(1-s)\leqslant 1$, we have $\bar{\beta}\leqslant 1$. Substituting Eqs. (2.19) and (2.21) into Eq. (2.18), we get

$$\begin{aligned}\dot{V}_1\leqslant&\frac{\xi_1^2\Gamma_1^2\theta^*\|S_1(x_1)\|^2}{2a_1^2}+\frac{1}{2}a_1^2+\frac{\xi_1^2\Gamma_1^2}{2}+\frac{1}{2}\bar{\varepsilon}_1^2+\frac{3\xi_1^2\Gamma_1^2\bar{\beta}^2}{2}\\ &+\frac{\lambda_2^2}{2}+\frac{z_2^2}{2}+\frac{1}{2}\|e\|^2+\xi_1\Gamma_1\beta\alpha_1-\xi_1\Gamma_1\left(\dot{y}_{\mathrm{d}}+\frac{\tilde{x}_1\dot{\mu}}{\mu}\right)-\frac{1}{r_0}\tilde{\theta}\dot{\hat{\theta}}\end{aligned} \tag{2.22}$$

The Nussbaum function chosen in this chapter is $N(\psi)=\psi^2\sin\psi$. We design the following virtual control signal α_1.

$$\begin{aligned}\alpha_1 =& N(\psi)\left(c_1\xi_1+\frac{\xi_1\Gamma_1^2}{2}+\frac{\xi_1\Gamma_1^2\hat{\theta}\|S_1(x_1)\|^2}{2a_1^2}\right.\\ &\left.+\frac{3\xi_1\Gamma_1^2\overline{\beta}^2}{2}-\Gamma_1\left(\dot{y}_{\mathrm{d}}+\frac{x_1\dot{\mu}}{\mu}\right)\right)\end{aligned} \tag{2.23}$$

$$\dot{\psi}=\xi_1\left(c_1\xi_1+\frac{\xi_1\Gamma_1^2}{2}+\frac{\xi_1\Gamma_1^2\hat{\theta}\|S_1(x_1)\|^2}{2a_1^2}+\frac{3\xi_1\Gamma_1^2\overline{\beta}^2}{2}-\Gamma_1\left(\dot{y}_{\mathrm{d}}+\frac{x_1\dot{\mu}}{\mu}\right)\right).$$

Substituting Eq. (2.23) into Eq. (2.22), we get

$$\begin{aligned}\dot{V}_1\leqslant&-c_1\xi_1^2+(\Gamma_1\beta N(\psi)+1)\dot{\psi}+\frac{1}{r_0}\tilde{\theta}\left(\frac{r_0\xi_1^2\Gamma_1^2S_1(x_1)^2}{2a_1^2}-\dot{\hat{\theta}}\right)\\ &+\frac{\bar{\varepsilon}_1^2}{2}+\frac{\lambda_2^2}{2}+\frac{z_2^2}{2}+\frac{a_1^2}{2}+\frac{\|e\|^2}{2}.\end{aligned} \tag{2.24}$$

Step $m(2 \leqslant m \leqslant n-1)$: Choose the Lyapunov function for the mth step

$$V_m = V_{m-1} + \frac{z_m^2}{2}. \tag{2.25}$$

Differentiating V_m with respect to time and using Eq. (2.16), we obtain

$$\begin{aligned}\dot{V}_m = & \ \dot{V}_{m-1} + z_m Q_m \Big(G_m + z_{m+1} + \lambda_{m+1} + \alpha_m + l_m e_1 \Big) - z_m \dot{\omega}_m \\ \leqslant & \ \dot{V}_{m-1} + z_m Q_m G_m + \frac{3}{2} z_m^2 Q_m^2 + \frac{z_{m+1}^2}{2} + \frac{\lambda_{m+1}^2}{2} + \frac{l_m^2}{2} \|e\|^2 \\ & - z_m \dot{\omega}_m + z_m Q_m \alpha_m .\end{aligned} \tag{2.26}$$

Applying Young's inequality and Lemma 2.1, we derive

$$z_m Q_m G_m \leqslant \frac{z_m^2 Q_m^2 \theta^* \left\| S_m \left(\bar{s}_m \right) \right\|^2}{2 a_m^2} + \frac{1}{2} a_m^2 + \frac{z_m^2 Q_m^2}{2} + \frac{1}{2} \bar{\varepsilon}_m^2, \tag{2.27}$$

where $\bar{s}_m = [s_2, \ldots, s_m]^{\mathrm{T}}$. We design the virtual control signal α_m as follows

$$\alpha_m = \frac{1}{Q_m} \left(-c_m z_m - \frac{z_m Q_m^2 \hat{\theta} \left\| S_m \left(\bar{s}_m \right) \right\|^2}{2 a_m^2} - 2 z_m Q_m^2 - \frac{z_m}{2} + \dot{\omega}_m \right). \tag{2.28}$$

Substituting Eqs. (2.24), (2.27) and (2.28) into Eq. (2.26), we get

$$\begin{aligned}\dot{V}_m \leqslant & \left(\Gamma_1 \beta N(\psi) + 1 \right) \dot{\psi} - \sum_{i=1}^{m} c_i z_i^2 - c_1 \xi_1^2 + \frac{1}{2} \sum_{i=1}^{m} \left(a_i^2 + \lambda_{i+1}^2 + \bar{\varepsilon}_i^2 \right) \\ & + \frac{1}{r_0} \tilde{\theta} \left(\frac{r_0 \xi_1^2 \Gamma_1^2 \left\| S_1 \left(x_1 \right) \right\|^2}{2 a_1^2} + \sum_{i=2}^{m} \frac{r_0 z_i^2 Q_i^2 \left\| S_i \left(\bar{s}_i \right) \right\|^2}{2 a_i^2} - \dot{\hat{\theta}} \right) \\ & + \left(\frac{1}{2} + \sum_{i=2}^{m} \frac{l_i^2}{2} \right) \|e\|^2 + \frac{z_{m+1}^2}{2}.\end{aligned} \tag{2.29}$$

Step n : Choose the Lyapunov function for the nth step

$$V_n = V_{n-1} + \frac{z_n^2}{2}. \tag{2.30}$$

The time derivative of V_n is

$$\begin{aligned}\dot{V}_n &= \dot{V}_{n-1} + z_n Q_n \left(u + l_n e_1 \right) - z_n \dot{\omega}_n \\ &\leqslant \dot{V}_{n-1} + \frac{z_n^2 Q_n^2}{2} + \frac{l_n^2}{2} \|e\|^2 - z_n \dot{\omega}_n + z_n Q_n u.\end{aligned} \tag{2.31}$$

We design the controller u and the adaptive law $\dot{\hat{\theta}}$ as follows

$$u = \frac{1}{Q_n}\left(-c_n z_n - \frac{z_n Q_n^2}{2} - \frac{z_n}{2} + \dot{\omega}_n\right) \tag{2.32}$$

$$\dot{\hat{\theta}} = \sum_{i=2}^{n-1} \frac{r_0 z_i^2 Q_i^2 \left\|S_i\left(\overline{s}_i\right)\right\|^2}{2a_i^2} - \sigma\hat{\theta} + \frac{r_0 \xi_1^2 \Gamma_1^2 \left\|S_1\left(x_1\right)\right\|^2}{2a_1^2}. \tag{2.33}$$

Substituting Eqs. (2.29), (2.32) and (2.33) into Eq. (2.31), we obtain

$$\begin{aligned} \dot{V}_n \leqslant & \left(\Gamma_1 \beta N(\psi) + 1\right)\dot{\psi} - c_1\xi_1^2 - \sum_{i=2}^{n} c_i z_i^2 + \frac{1}{2}\sum_{i=1}^{n}(a_i^2 + \bar{\varepsilon}_i^2) \\ & + \frac{1}{2}\sum_{i=2}^{n}\lambda_i^2 + \left(\frac{1}{2} + \sum_{i=2}^{n}\frac{l_i^2}{2}\right)\|e\|^2 + \frac{\sigma}{r_0}\tilde{\theta}\hat{\theta}. \end{aligned} \tag{2.34}$$

Based on V_e and V_n, we define the Lyapunov function as

$$V = V_e + V_n + \frac{1}{2}\sum_{i=2}^{n}\lambda_i^2. \tag{2.35}$$

Differentiating V with respect to time, we get

$$\dot{V} = \dot{V}_e + \dot{V}_n + \sum_{i=2}^{n}\lambda_i\left(-\frac{\lambda_i}{\tau_i} + M_i(\cdot)\right). \tag{2.36}$$

Using Eqs. (2.34) and (2.36), and applying Young's inequality, we derive

$$\frac{\sigma}{r_0}\tilde{\theta}\hat{\theta} \leqslant \frac{\sigma}{2r_0}\left(\theta^{*2} - \tilde{\theta}^2\right) \tag{2.37}$$

$$\sum_{i=2}^{n}\lambda_i M_i(\cdot) \leqslant \sum_{i=2}^{n}\left(\frac{\lambda_i^2 M_i^2(\cdot)}{2} + \frac{1}{2}\right). \tag{2.38}$$

Substituting Eqs. (2.11), (2.34), (2.37) and (2.38) into Eq. (2.36), and rearranging the terms, we get

$$
\begin{aligned}
\dot{V} \leqslant & -\left(\upsilon - 2\eta - \frac{1}{\varrho} - \frac{1}{2} - \sum_{i=2}^{n} \frac{l_i^2}{2}\right) e^{\mathrm{T}} e + \eta^{-1} \|P\|^2 \theta^* + \varrho \|P\|^2 \bar{\varpi}^2 \\
& + \eta^{-1} \bar{\varepsilon}_0^2 + (\Gamma_1 \beta N(\psi) + 1)\, \dot{\psi} + \frac{1}{2} \sum_{i=1}^{n} \left(a_i^2 + \bar{\varepsilon}_i^2\right) - c_1 \xi_1^2 \\
& + \sum_{i=2}^{n} \left(-\frac{1}{\tau_i} + \frac{1}{2} + \frac{M_i^2(\cdot)}{2}\right) \lambda_i^2 - \sum_{i=2}^{n} c_i z_i^2 - \frac{\sigma}{2r_0} \tilde{\theta}^2 + \frac{\sigma}{2r_0} \theta^{*2} + \frac{n-1}{2}.
\end{aligned} \tag{2.39}
$$

We define the following parameters

$$
\begin{aligned}
q = & \min \left\{2c_i,\ \sigma,\ \frac{2}{\tau_i} - 1 - M_i^2(\cdot)\right\} \\
\tilde{b} = & \eta^{-1} \|P\|^2 \theta^* + \varrho \|P\|^2 \bar{\varpi}^2 + \frac{1}{2} \sum_{i=1}^{n} \left(a_i^2 + \bar{\varepsilon}_i^2\right) + \frac{\sigma}{2r_0} \theta^{*2} + \frac{n-1}{2} + \eta^{-1} \bar{\varepsilon}_0^2.
\end{aligned}
$$

There exists a parameter υ that satisfies the following inequality

$$
\upsilon > 2\eta + \frac{1}{\varrho} + \frac{1}{2} + \sum_{i=2}^{n} \frac{l_i^2}{2} + q \lambda_{\max}(P),
$$

where $\lambda_{\max}(P)$ is the maximum eigenvalue of matrix P. Then Eq. (2.39) can be rewritten as

$$
\dot{V}(t) \leqslant -qV(t) + (\Gamma_1 \beta N(\psi) + 1)\, \dot{\psi} + \tilde{b}. \tag{2.40}
$$

According to Lemma 2.2, $(\Gamma_1 \beta N(\psi) + 1)\, \dot{\psi}$ is bounded. Therefore, we let $(\Gamma_1 \beta N(\psi) + 1)\, \dot{\psi} + \tilde{b} \leqslant D$, where D is a positive constant. Substituting D into Eq. (2.40), we obtain

$$
V(t) \leqslant \frac{D}{q} + \left[V(0) - \frac{D}{q}\right] \mathrm{e}^{-qt}. \tag{2.41}
$$

The above equation shows that all signals are semi-global uniformly ultimately bounded. Let $\lambda_{\min}(P)$ be the minimum eigenvalue of matrix P, then from Eq. (2.41), we have

$$
|\xi_1| \leqslant \sqrt{\frac{2D}{q} + 2\left[V(0) - \frac{D}{q}\right] \mathrm{e}^{-qt}} \tag{2.42}
$$

$$|e_i| \leqslant \|e\| \leqslant \sqrt{\frac{\frac{D}{q} + \left(V(0) - \frac{D}{q}\right)\mathrm{e}^{-qt}}{\lambda_{\min}(P)}}. \tag{2.43}$$

Define $\Delta = \sqrt{\frac{\frac{D}{q} + \left(V(0) - \frac{D}{q}\right)\mathrm{e}^{-qt}}{\lambda_{\min}(P)}}$. Since $x_1 = \tilde{x}_1 + y_\mathrm{d}$, and $y_\mathrm{d} \leqslant Y_0$, we have $|x_1| \leqslant |\tilde{x}_1| + |y_\mathrm{d}|$. Therefore, $-\mu(t) - Y_0 < x_1 < \mu(t) + Y_0$. Defining $\underline{k}_{c1} = \bar{k}_{c1} = \mu(0) + Y_0$, then one get $-\underline{k}_{c1} < x_1 < \bar{k}_{c1}$. From $x_m = e_m + \beta\hat{\chi}_m$, $|x_m| \leqslant |e_m| + |\beta\hat{\chi}_m|$, and $\underline{k}_{cm} = \Delta + \bar{\beta}\underline{K}_m$, $\bar{k}_{cm} = \Delta + \bar{\beta}\bar{K}_m$, then inequality $-\underline{k}_{cm} < x_m < \underline{k}_{cm} (m = 2, \ldots, n)$ holds.

Define $\phi = \sqrt{\frac{2D}{q} + 2\left(V(0) - \frac{D}{q}\right)\mathrm{e}^{-qt}}$. From Eqs. (2.13) and (2.42), it can be seen that

$$\begin{aligned} \frac{\tilde{x}_1^2}{\mu^2 - \tilde{x}_1^2} = \xi_1^2 &\leqslant \frac{2D}{q} + 2\left(V(0) - \frac{D}{q}\right)\mathrm{e}^{-qt} \\ \tilde{x}_1^2 &\leqslant \frac{\phi^2\mu^2}{1+\phi^2} \\ |\tilde{x}_1| &\leqslant \sqrt{\frac{\phi^2}{1+\phi^2}}|\mu| < |\mu|. \end{aligned} \tag{2.44}$$

The above formula demonstrates that, by applying the specified performance control design, the system tracking error can be always bounded within the range of the specified performance function, and the proper parameter selection can make the system tracking error approach zero.

Remark 2.4 From formula (2.41), it can be observed that under the framework of backstepping method, based on the constructed Lyapunov function, the derived control law u can ensure the boundedness of all error signals $z_m (m = 2, \ldots, n)$, and guarantee that the tracking error x_1 satisfies the preset performance specifications.

In summary, for the original system (2.1), using the designed virtual control laws (2.23), (2.28) and control law (2.32), we can ensure the boundedness of all states of the original system, the fulfillment of the set performance specifications for the system tracking error, and the preservation of good transient performance and steady-state performance in the event of actuator failure.

2.4 Simulation Analysis

Consider the following pure feedback nonlinear system

$$\begin{cases} \dot{x}_1 = x_2 + 0.5x_2 \sin^2 x_1 \\ \dot{x}_2 = u_\mathrm{f} - 0.5x_1x_2^3 \\ y = x_1. \end{cases}$$

The system tracks the trajectory of $y_\mathrm{d} = \cos(0.4t)$. The specified performance function is $\mu = \ln\left(0.2294(2-t)^{1.5} + 1\right) + 0.05$. The system state constraint boundaries are $\underline{k}_{\mathrm{c}1} = \bar{k}_{\mathrm{c}1} = 1.55,\ \underline{k}_{\mathrm{c}2} = \bar{k}_{\mathrm{c}2} = 1.6$. The parameters are set as $c_1 = c_2 = 5,\ \sigma_1 = 0.02,\ a_1 = 10,\ \tau_1 = 0.2,\ \omega_2(0) = 0.2,\ \bar{K}_2 = \underline{K}_2 = 1.1,\ \bar{\beta} = 1,\ \varsigma = 0.5,\ \bar{n} = 1.5,\ \mu_\infty = 0.05,\ t_0 = 0,\ t_\mathrm{f} = 2,\ l_1 = l_2 = 10,\ C_0 = 0.2294$.

We consider the system actuator fault

$$u_\mathrm{f} = b(1-s)u + \varpi,$$

where the parameters are set as $b = 1,\ s = 0.05$, and the actuator deviation fault nonlinear function is $\varpi = 0.1\cos\left(-t^2\right)$. The initial values of the system adaptive parameters are set as $x_1(0) = 0.8,\ x_2(0) = 0.6,\ \hat{\theta}_{10} = 0,\ r_0 = 80,\ \hat{\chi}_{10} = \hat{\chi}_{20} = 1,\ \psi(0) = 3$.

Figures 2.1, 2.2, 2.3, 2.4, 2.5, 2.6 and 2.7 present the simulation results of this chapter. Figure 2.1 demonstrates that the system output tracks the target function well under this control method, and the system states $x_1 = y$ and x_2 remain within the constraint range. In contrast, Fig. 2.2 indicates that without compensating for the system actuator fault (let $N(\psi) = 1$, other design parameters unchanged), the system input exhibits severe jitter, the system becomes unstable, and cannot continue to run after 0.27 s of simulation, when the system tracking error $\tilde{x}_1$ approaches the performance function boundary. This suggests that the control method of this chapter can effectively compensate for the actuator fault that occurs in the system, and can keep the system running stably. Figure 2.3 shows that when the system state is not constrained, the system state x_2 significantly exceeds the constraint condition in the early stage of simulation. This indicates that the control method of this chapter can keep the system state within the constraint range and meet the state constraint requirements of the system. Figures 2.4 and 2.5 illustrate, respectively, the convergence situation of system tracking error when performance function takes different t_f and $\bar{n}$. It can be seen that under the performance function used in this chapter, it can ensure that the system tracking error converges within time t_f, and can change the error convergence speed by setting $\bar{n}$, $\bar{n}$ is larger, and the error convergence speed is faster. However, with the decrease of parameter t_f or increase of $\bar{n}$, the frequency of system state oscillation will also increase accordingly. Figure 2.6 depicts the tracking situation of system when performance function is $\mu^* = (\mu_0 - \mu_\infty)\,\mathrm{e}^{-\bar{m}t} + \mu_\infty$, ($\mu_0 = 0.55,\ \mu_\infty = 0.05,\ \bar{m} = 1$), where μ^* is the performance function form used in Yang et al. (2019), Zhao et al. (2021). The control effect and system stability time

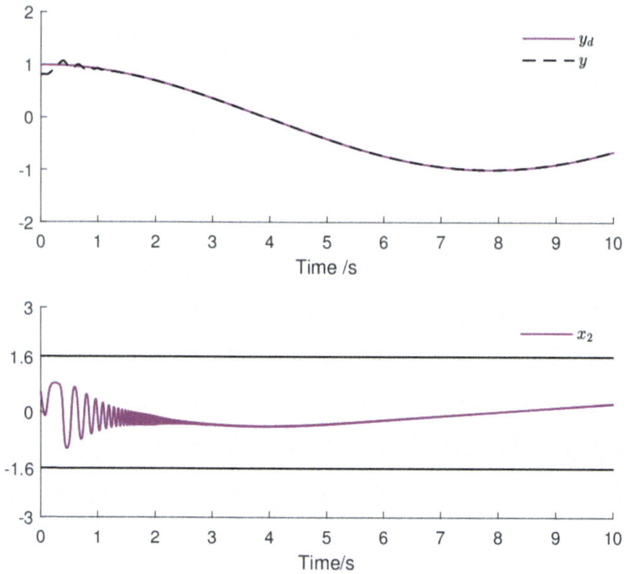

Fig. 2.1 Tracking signal y_d, output y and system state x_2

Fig. 2.2 Tracking error $\tilde{x}_1$ and system input u without actuator fault compensation

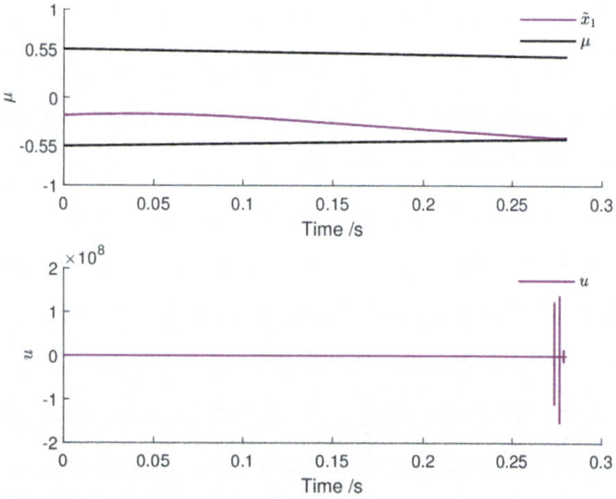

of this algorithm depend on the parameter settings of μ_0 , μ_∞ and $\bar{m}$, so it cannot flexibly set the convergence speed and convergence time of system tracking error. Therefore, the performance function used in this chapter is more flexible than the performance function μ^* used in Kostarigka and Rovithakis (2012), Ni et al. (2019). Figure 2.7 shows the control input of this chapter.

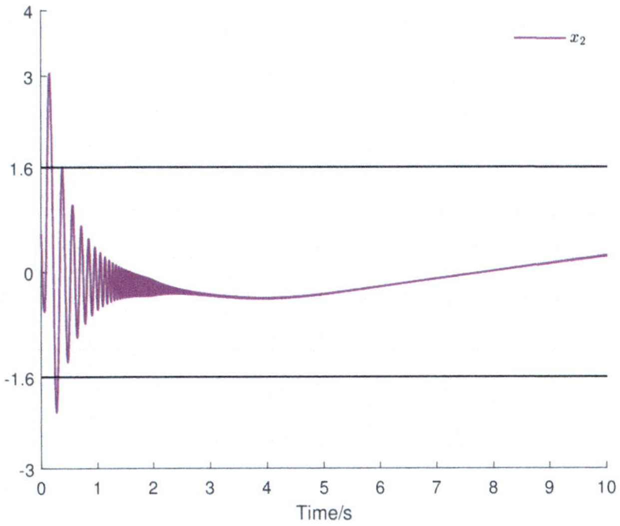

Fig. 2.3 System state without constraints x_2

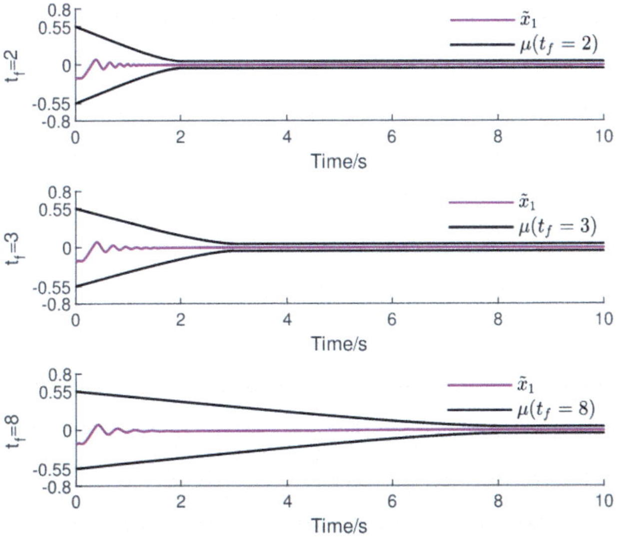

Fig. 2.4 Tracking error $\tilde{x}_1$ with different convergence time t_f, $(\bar{n} = 1.5)$

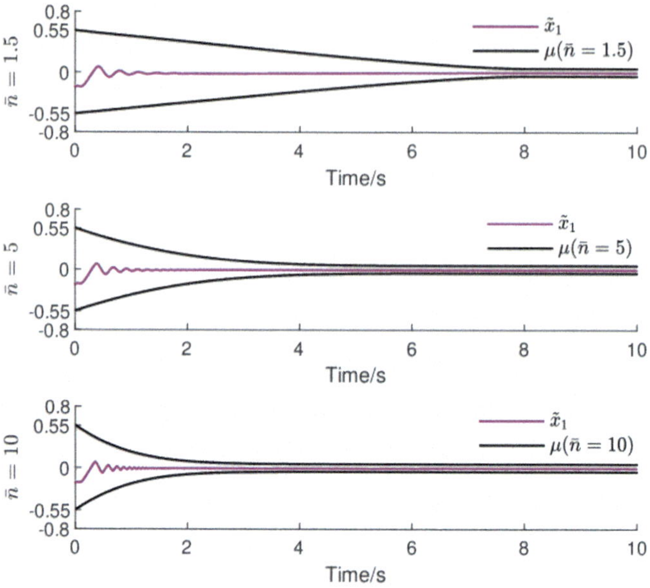

Fig. 2.5 Tracking error $\tilde{x}_1$ with different convergence speed $\bar{n}$, ($t_f = 8$)

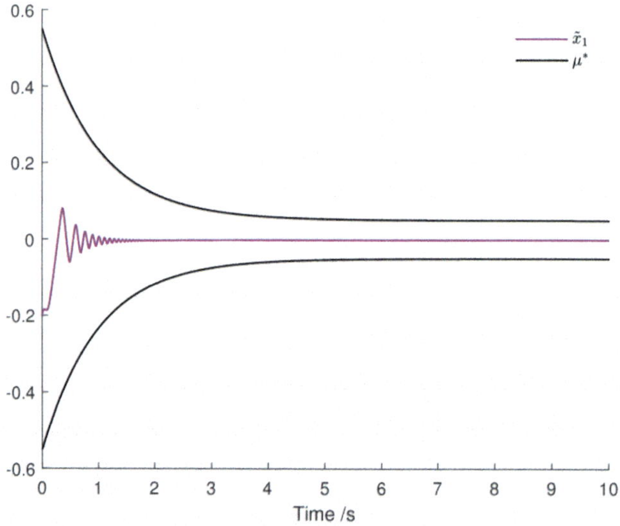

Fig. 2.6 Tracking error $\tilde{x}_1$ with performance function μ^*

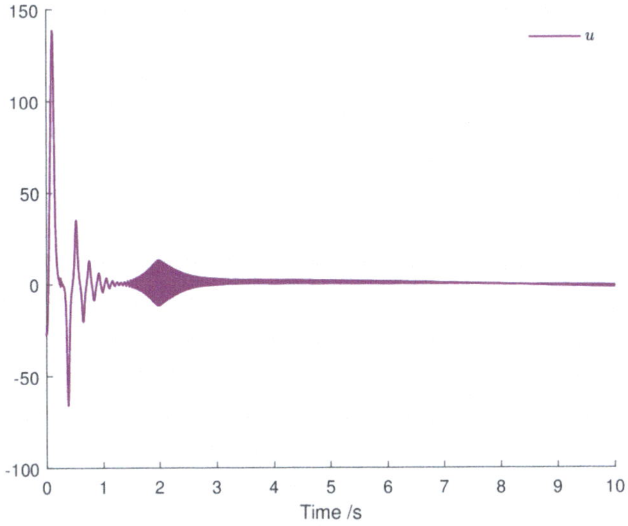

Fig. 2.7 Control input u

2.5 Conclusion

This chapter proposes an adaptive neural network output-feedback fault-tolerant control method for a class of state-unmeasurable pure feedback systems with actuator faults, considering the constraint conditions of full state constraints and specified performance. First, a state observer is established to deal with the problem of state unmeasurability of the system. By introducing a nonlinear mapping, the state-constrained system is transformed into a new system without constraints, thus solving the problem of full state constraints of the system. Second, radial basis neural networks are used to approximate the unknown nonlinear functions in the system. In addition, this chapter adopts a new performance function, which effectively controls the convergence speed of tracking error and ensures that the error converges within the set time. Finally, simulation results verify the effectiveness of the control algorithm proposed in this chapter.

References

Arefi MM, Zarei J, Karimi HR (2014) Adaptive output feedback neural network control of uncertain non-affine systems with unknown control direction. J Franklin Inst 351(8):4302–4316

Boulkroune A (2016) A fuzzy adaptive control approach for nonlinear systems with unknown control gain sign. Neurocomputing 179:318–325

Kostarigka AK, Rovithakis GA (2012) Adaptive dynamic output feedback neural network control of uncertain MIMO nonlinear systems with prescribed performance. IEEE Trans Neural Netw Learn Syst 23(1):138–149

Li Y, Tong S (2017) Adaptive neural networks decentralized FTC design for nonstrict-feedback nonlinear interconnected large-scale systems against actuator faults. IEEE Trans Neural Netw Learn Syst 28(11):2541–2554

Li Y, Tong S (2018) Adaptive neural networks prescribed performance control design for switched interconnected uncertain nonlinear systems. IEEE Trans Neural Netw Learn Syst 29(7):3059–3068

Li DP, Liu YJ, Tong S, Chen CP, Li DJ (2019) Neural networks-based adaptive control for nonlinear state constrained systems with input delay. IEEE Trans Cybernet 49(4):1249–1258

Lin G, Li H, Ma H, Yao D, Lu R (2022) Human-in-the-loop consensus control for nonlinear multi-agent systems with actuator faults. IEEE/CAA J Automatica Sinica 9(1):111–122

Liu C, Wang H, Liu X, Zhou Y (2021) Adaptive finite-time fuzzy funnel control for nonaffine nonlinear systems. IEEE Trans Syst Man Cybernet Syst 51(5):2894–2903

Li H, Wu Y, Chen M (2021) Adaptive fault-tolerant tracking control for discrete-time multiagent systems via reinforcement learning algorithm. IEEE Trans Cybernet 51(3):1163–1174

Ma H, Li H, Liang H, Dong G (2019) Adaptive fuzzy event-triggered control for stochastic nonlinear systems with full state constraints and actuator faults. IEEE Trans Fuzzy Syst 27(11):2242–2254

Ma H, Li H, Lu R, Huang T (2020) Adaptive event-triggered control for a class of nonlinear systems with periodic disturbances. Sci China-Inf Sci 63(5):150212

Mao J, Zhang T, Xia X, Shen Q, Yi Y (2016) Adaptive output feedback control for multi-input single-output systems with actuator failures. Control Theory Appl 33(4):512–522

Ni J, Ahn CK, Liu L, Liu C (2019) Prescribed performance fixed-time recurrent neural network control for uncertain nonlinear systems. Neurocomputing 363:351–365

Ren B, Ge SS, Tee KP, Lee TH (2010) Adaptive neural control for output feedback nonlinear systems using a barrier Lyapunov function. IEEE Trans Neural Netw 21(8):1339–1345

Ren H, Karimi HR, Lu R, Wu Y (2020) Synchronization of network systems via aperiodic sampled-data control with constant delay and application to unmanned ground vehicles. IEEE Trans Indus Electron 67(6):4980–4990

Tian W, Wang Y, Li Y (2015) Fuzzy adaptive output tracking control for a class of uncertain nonlinear systems. J Guangdong Univ Technol 32(1):91–97

Wang CC, Yang GH (2018) Observer-based adaptive prescribed performance tracking control for nonlinear systems with unknown control direction and input saturation. Neurocomputing 284:17–26

Wang H, Chen B, Lin C, Sun Y (2016) Observer-based adaptive neural control for a class of nonlinear pure-feedback systems. Neurocomputing 171:1517–1523

Wang F, Chen B, Lin C, Zhang J, Meng X (2018) Adaptive neural network finite-time output feedback control of quantized nonlinear systems. IEEE Trans Cybernet 48(6):1839–1848

Yang H, Shi P, Zhao X, Shi Y (2016) Adaptive output-feedback neural tracking control for a class of nonstrict-feedback nonlinear systems. Inf Sci 334:205–218

Yang B, Zhou Q, Cao L, Lu R (2019) Event-triggered control for multi-agent systems with prescribed performance and full state constraints. Acta Automatica Sinica 45(8):1527–1535

Zeng G, Sun B, Liang H (2000) A direct adaptive controller based on neural network. J Guangdong Univ Technol 17(1):20–23

Zhang JX, Yang GH (2017) Prescribed performance fault-tolerant control of uncertain nonlinear systems with unknown control directions. IEEE Trans Autom Control 62(12):6529–6535

Zhang T, Xia M, Yi Y (2017) Adaptive neural dynamic surface control of strict-feedback nonlinear systems with full state constraints and unmodeled dynamics. Automatica 81:232–239

Zhao G, Gao R, Chen J (2021) Adaptive prescribed performance control of quadrotor with unknown actuator fault. Control Decision 36(9):2103–2112

Zheng X, Dong G, Zhou Q, Lu R (2020) Fault-tolerant control for stochastic multi-agent systems with output constraints. Control Theory Appl 37(5):961–968

Zhou Q, Shi P, Xu S, Li H (2013) Observer-based adaptive neural network control for nonlinear stochastic systems with time delay. IEEE Trans Neural Netw Learn Syst 24(1):71–80

Zhou Q, Zhao S, Li H, Lu R, Wu C (2019) Adaptive neural network tracking control for robotic manipulators with dead zone. IEEE Trans Neural Netw Learn Syst 30(12):3611–3620

Zhou Q, Chen G, Lu R, Bai W (2019) Disturbance-observer-based event-triggered control for multi-agent systems with input saturation. Scientia Sinica Informationis 49(11):1502–1516

Chapter 3
Approximation-Based Nussbaum Gain Adaptive Control with Periodic Disturbances

Abstract This chapter considers the Nussbaum gain adaptive control issue for a type of nonlinear systems, in which some sophisticated and challenging problems, such as periodic disturbances, dead zone output, and unknown control direction are addressed. The Fourier series expansion and radial basis function neural network are incorporated into a function approximator to model time-varying-disturbed function with a known period in nonlinear systems. To deal with the problems of the dead zone output and unknown control direction, the Nussbaum-type function is recommended in the design of the control algorithm. Applying the Lyapunov stability theory and backstepping technique, the proposed control strategy ensures that the tracking error is pulled back to a small neighborhood of origin and all closed-loop signals are bounded. Finally, simulation results are presented to show the availability and validity of the analysis approach.

3.1 Introduction

The adaptive backstepping method, as a practical technique, has attracted a great attention for controlling systems, and a lot of crucial progress have been made during the past decades (Krstić et al. 1992; Sokolov et al. 2015; Xia et al. 2018; Zeng and Wang 2008; Liu and Li 2019; Liu and Wang 2015; Zhu et al. 2017; Hu et al. 2021). As a recursive design process, backstepping was integrated into the adaptive controller design by Kanellakopoulos et al. (1991) for a category of feedback linearizable systems. Subsequently, the backstepping-based adaptive approach was developed further to analyze a variety of nonlinear systems, such as chaotic systems (Ge et al. 2000), multi-agent systems (Roy et al. 2018; Lin et al. 2022; Liu et al. 2021), and complex-constrained nonlinear systems (Zhang et al. 2020, 2016; Hu et al. 2021; Liu et al. 2017; Huang et al. 2021; Bai et al. 2020; Zhao et al. 2020). Combining the adaptive backstepping technique and the impedance learning method, the adaptive fuzzy neural network (NN) control problem was investigated in He and Dong (2018) for an uncertain constrained robot, in which a tan-type barrier Lyapunov function was used to handle the influence of state constraints. Followed by this, with the asymmetric time-variant barrier Lyapunov function, an adaptive backstepping control

H. Ren et al., *Intelligent Control of Nonlinear Systems and Nonlinear Multi-Agent Systems with Complex Constraints*, Intelligent Control and Learning Systems 16,
https://doi.org/10.1007/978-981-96-0320-6_3

approach was proposed in Wang et al. (2020) for a type of saturated nonlinear systems subject to partial state constraints.

To be exact, the adaptive backstepping control issue for nonlinear systems against periodic disturbances is complicated and challenging, and it has accepted considerable attention (Zhang and Zhang 2016; Muramatsu and Katsura 2018). For a type of nonlinear systems subject to time-varying periodic disturbances, an adaptive control strategy was developed in Chen (2009), in which NNs and the Fourier series expansion (FSE) were incorporated into a function approximator to approximate unknown disturbed functions. Moreover, Chen et al. (2010) combined the fuzzy logic systems (FLSs) and FSE to ensure the stability of nonlinear systems, in which the unknown nonlinear functions contained unmeasured time-dependent periodic disturbances. Followed by Chen (2009), the control method was extended to a category of nonlinear multiple-input and multiple-output (MIMO) systems (Zuo et al. 2019), and the bounded condition of unknown nonaffine function has been relaxed.

The Nussbaum gain method, which was brought forward by Nussbaum (1983), is an effective tool, and it paves the way to handle the unknown sign problems. Based on this, many important achievements have been published in recent years (Ge and Wang 2003; Chen et al. 2017; Tong et al. 2020; Wang et al. 2020; Liang et al. 2020; Wu et al. 2020). To obtain the objective that unknown control direction does not influence the control system performance, a Nussbaum-type function was utilized in the controller design (Wang et al. 2016). Considering a type of nonlinear MIMO systems against nonparametric uncertainties, a robust fault-tolerant adaptive PID control method was developed in Song et al. (2017) by employing the Nussbaum gain technique. In addition, we found an interesting thing, that is, the Nussbaum-type function can also be used to address dead zone output problems. For instance, by introducing a Nussbaum-type function, the unknown coefficient difficulty resulting from the derivation in the system output mechanism was handled in Wang et al. (2015). To dispose with the issue of dead zone output, the Nussbaum-type function was used in the work of Liu et al. (2015) for a type of nonlinear systems against time delay. Nevertheless, for a category of nonlinear systems against periodic disturbances, dead zone output, and unknown control direction, the issue of backstepping adaptive control has not been fully studied, which is a challenging and interesting task.

Based on the above discussion, this article provides a Nussbaum gain NN adaptive control strategy for a category of nonlinear systems with dead zone output, unknown control direction, and periodic disturbances. The primary contributions of this chapter are summarized as follows.

1. Different from the common nonlinear strict-feedback systems, the considered control system model suffers from the periodic disturbances, dead zone output, and unknown control direction at the same time. Without a doubt, it improves the difficulty of the control algorithm design. But the proposed control approach in this article still guarantees high quality tracking performance.
2. Unlike the existing results (Ma et al. 2019; Liu and Tong 2017), the nonlinear functions $f_i(\cdot)$ $(i = 1, 2, \ldots, n)$ in the system (3.1) are unknown. To identify the

unknown nonlinearities, the radial basis function neural network (RBFNN) and FSE are considered in this chapter.

3. Compared with the previous results (Liu et al. 2018; Dong et al. 2021), for the system (3.1), the main characteristic is that the unknown nonlinear functions $f_i(\cdot)$ $(i = 1, 2, \ldots, n)$ include the periodically time-variant disturbances $d_i(t)$ $(i = 1, 2, \ldots, n)$, which approximated by FSE in this article. Moreover, the Nussbaum-type function is applied to handle dead zone output and unknown control direction problems. In the end, simulation results are given to illustrate the availability and correctness of the designed NN control strategy.

Notations: The real n-dimensional space is represented by $\mathbb{R}^n$. For a given matrix or vector A, A^T denotes the transpose. $|C|$ is the absolute value, and $||C||$ represents the Euclidean norm. $||C||_F$ stands for the Frobenius norm.

3.2 Preliminaries

Consider the following nonlinear system with periodic disturbances and dead zone output as

$$\begin{aligned} \dot{x}_i &= a_i x_{i+1} + f_i(\bar{x}_i, d_i(t)) \\ \dot{x}_n &= a_n u + f_n(x, d_n(t)) \\ y &= \psi(x_1), \; i = 1, 2, \ldots, n-1, \end{aligned} \tag{3.1}$$

where $u \in \mathbb{R}$ and $y \in \mathbb{R}$ represent the control input and nonlinear system output, respectively. $\bar{x}_i = [x_1, \ldots, x_i]^T \in \mathbb{R}^i$ indicates the system state vector. Particularly, $x = \bar{x}_n$. a_i $(i = 1, 2, \ldots, n)$ denote unknown sign constants and $d_i(t) : [0, +\infty) \to \mathbb{R}^{m_i}$ $(i = 1, 2, \ldots, n)$ show the continuous but time-varying unknown disturbances with known periods T_i, i.e., $d_i(t + T_i) = d_i(t)$. In addition, the unknown and continuous nonlinear functions are represented by $f_i(\bar{x}_i, d_i(t)) : \mathbb{R}^{i+m_i} \to \mathbb{R}, i = 1, 2, \ldots, n$.

In Wang et al. (2015), the dead zone output $\psi(x_1)$ is described by

$$y = \psi(x_1) = \begin{cases} \bar{m}_l(x_1 - c_l), & x_1 \le c_l \\ 0, & c_l < x_1 < c_r \\ \bar{m}_r(x_1 - c_r), & x_1 \ge c_r \end{cases} \tag{3.2}$$

with $\bar{m}_l > 0$ and $\bar{m}_r > 0$. Moreover, $c_r > 0$ and $c_l < 0$ denote the breakpoints of the output nonlinearity. In the process of backstepping design method, we need to take the derivative of the dead zone output. However, in the non-smooth dead zone model (3.2), the dead zone output $\psi(x_1)$ is not differentiable at the points of sharp corner $x_1 = c_r$ and $x_1 = c_l$. To overcome this problem, the dead zone output model is further formulated as

$$x_1 = \psi^{-1}(y) = \frac{\bar{m}_r c_r + y}{\bar{m}_r}\beta_{r0}(y) + \frac{\bar{m}_l c_l + y}{\bar{m}_l}\beta_{l0}(y). \tag{3.3}$$

For the adjusted positive parameter C_0, design the adjustable functions $\beta_{r0}(y)$ and $\beta_{l0}(y)$ as

$$\beta_{r0}(y) = \frac{2\arctan(C_0 y) + \pi}{2\pi} \tag{3.4}$$

$$\beta_{l0}(y) = \frac{-2\arctan(C_0 y) + \pi}{2\pi}. \tag{3.5}$$

The further proposed model can approximate the non-smooth dead zone model (3.2) with any accuracy by adjusting parameter.

Remark 3.1 It should be emphasized that there exist two kinds of periodic disturbances, one is time-dependent periodic disturbances and the other is state-dependent periodic disturbances. The time-dependent periodic disturbances can be denoted as $d_i(t)$ $(i = 1, 2, \ldots, n)$, which satisfying $d_i(t + T_i) = d_i(t)$. The state-dependent periodic disturbances can be shown as $d_i(x)$ with $d_i(x + T_i) = d_i(x)$ $(i = 1, 2, \ldots, n)$, where x is the system state and $T_i > 0$ stands for the periods. If the periodic disturbances in system (3.1) are state-dependent, then the unknown nonlinear functions $f_i(\cdot)$ $(i = 1, 2, \ldots, n)$ only depend on the system state $\bar{x}_i$. In this case, some existing adaptive NN control methods (Cao et al. 2019; Du et al. 2021) can be applied to handle this adaptive control problem for system (3.1) directly. Therefore, we will concentrate on the time-dependent periodic disturbances in this chapter.

Based on the above system descriptions, the following assumption, definition, and lemmas are given in this chapter.

Assumption 3.1 The signs of $a_i \neq 0$ $(i = 1, 2, \ldots, n)$ are unknown.

Definition 3.1 (Nussbaum 1983) The continuous function $N(\rho)$ has the following properties

$$\lim_{r\to\infty} \sup \frac{1}{r}\int_0^r N(\rho)d\rho = +\infty$$

$$\lim_{r\to\infty} \inf \frac{1}{r}\int_0^r N(\rho)d\rho = -\infty.$$

Then, $N(\rho)$ can be regarded as a Nussbaum-type function. Here we define $N(\rho) = \rho^2\cos(\rho)$.

Lemma 3.1 (Liu and Tong 2017) *For smooth functions $V(t) \geq 0$ and $\rho_i(t)$ $(i = 1, \ldots, n)$, which are defined on $[0, t_f]$, the parameters $D_1 > 0$, $D > 0$, and A_i $(i =$*

$1, 2, \ldots, n$), and the smooth Nussbaum-type function $N(\rho)$, the following expression holds

$$V(t) \leq D_1 + e^{-Dt} \sum_{i=1}^{n} \int_0^t A_i N(\rho_i)\dot{\rho}_i e^{D\tau} d\tau + e^{-Dt} \sum_{i=1}^{n} \int_0^t \dot{\rho}_i e^{D\tau} d\tau.$$

Then, $\rho_i(t)$, $V(t)$ and $\int_0^t N(\rho_i)\dot{\rho}_i d\tau$ are bounded in $[0, t_f]$.

Lemma 3.2 (Liu et al. 2015) *For the further proposed approximation dead zone output model* (3.3)–(3.5), *the following property holds*

$$\frac{dy}{dt} = \eta_1(t)\dot{x}_1$$

with the time-varying coefficient $\eta_1(t) = (dy/dx_1)$. $\eta_1(t)$ shows the slope of the dead zone output model and $0 < \eta_1(t) < C_M = \max\{\bar{m}_l, \bar{m}_r\}$.

3.3 FSE-RBFNN-Based Approximator Design

The major design obstacle of this chapter is that the periodically time-varying disturbances $d_i(t)$ $(i = 1, 2, \ldots, n)$ are not used as RBFNN inputs directly. At first, the periodic disturbances $d_i(t)$ $(i = 1, 2, \ldots, n)$ are estimated by employing FSE. Accordingly, the measured system variable χ_i and the estimated value of d_i are employed as the RBFNN inputs to approximate unknown nonlinear functions $F_i(\chi_i, d_i(t))$ $(i = 1, 2, \ldots, n)$.

For a measured variable $\chi \in \Omega_\chi \subset \mathbb{R}^l$ and a compact set Ω_χ, consider the unknown function $F(\chi, d(t))$ with a continuous but unknown disturbance vector $d(t) = [d_1(t), \ldots, d_m(t)]^T \in \Omega_d \subset \mathbb{R}^m$, where Ω_d is a compact set. Moreover, with the help of a linearly parameterized FSE, the period and continuous disturbance $d(t)$ can be described by

$$d(t) = d_0(t) + W^T G(t), \tag{3.6}$$

where $d_0(t)$ represents the truncation error and $\|d_0(t)\| \leq \bar{d}_0$. $\bar{d}_0$ is a positive parameter. $W = [\xi_1, \ldots, \xi_m] \in \mathbb{R}^{q \times m}$ and it is a constant matrix. Noting that $\xi_i \in \mathbb{R}^q$ consists of the first q parameters of the FSE of $d(t)$ with an odd number q and $i = 1, \ldots, m$. It is worth mentioning that the error would be made arbitrarily small by magnifying q and $G(t) = [g_1(t), \ldots, g_q(t)]^T$. The ith derivative of $G(t)$ is smooth and bounded with $i = 1, 2, \ldots, n$. For $j = 1, \ldots, (q-1)/2$, define

$$g_1(t) = 1$$
$$g_{2j}(t) = \sqrt{2}\sin(2\pi jt/T)$$
$$g_{2j+1}(t) = \sqrt{2}\cos(2\pi jt/T).$$

Using the RBFNN to approximate $F(\chi, d(t))$ over $\Omega = \Omega_\chi \times \Omega_d$, in which $d(t)$ is measured and Ω is a compact set. According to Park and Sandberg (1991), Ma et al. (2020), the unknown continuous function $F(\chi, d(t))$ is expressed as

$$F(\chi, d(t)) = \theta^T H(\chi, d(t)) + d_F(\chi, d(t)), \tag{3.7}$$

where $\theta = [\zeta_1, \ldots, \zeta_p]^T \in \mathbb{R}^p$ denotes the weight vector, and it is defined by

$$\theta := \arg\min_{\hat{\theta}\in\mathbb{R}^p}\{\sup_{(\chi, d(t))\in\Omega} |F(\chi, d(t)) - \hat{\theta}^T H(\chi, d(t))|\}.$$

$H(\chi, d(t)) = [h_1(\chi, d(t)), \ldots, h_p(\chi, d(t))]^T$ refers to a smooth vector-valued function with $h_j(\chi, d(t)) = \exp[-||z - \mu_j||^2/\upsilon^2]$ $(j = 1, \ldots, p)$, in which $z = [\chi^T, d^T(t)]^T$. $\mu_j \in \Omega$ represents the center of $h_j(\chi, d(t))$. Moreover, the positive real number υ stands for the width of $h_j(\chi, d(t))$. $d_F(\chi, d(t))$ stands for the inherent NN approximation error with $|d_F(\chi, d(t))| \le \bar{d}_F$, and $\bar{d}_F$ is a positive constant. Increasing the NN node number p leads to the decrease of the approximation error (Hu and Wang 2006; Liu et al. 2020, 2022; Zeng and Wang 2006; Zhou et al. 2019; Li et al. 2019, 2021; Zhu et al. 2020; Wu et al. 2021). Combining formula (3.6) and Eq. (3.7), one has

$$F(\chi, d(t)) = d_F(\chi, d(t)) + \theta^T H\left(\chi, W^T G(t) + d_0(t)\right).$$

Then, the FSE-RBFNN-based approximator is shown as $\mathcal{G}(\chi, t) = \theta^T H\left(\chi, W^T G(t)\right)$, which is utilized to approximate $F(\chi, d(t))$. Therefore, the following formula holds

$$F(\chi, d(t)) = \varepsilon(\chi, t) + \theta^T H\left(\chi, W^T G(t)\right)$$

with

$$\varepsilon(\chi, t) = d_F(\chi, d(t)) + \theta^T H\left(\chi, W^T G(t) + d_0(t)\right) - \theta^T H\left(\chi, W^T G(t)\right).$$

Lemma 3.3 (Chen 2009) *The approximation error $\varepsilon(\chi, t)$ satisfies*

$$|\varepsilon(\chi, t)| \le \bar{\varepsilon}$$

for $(\chi, d(t)) \in \Omega$, where $\bar{\varepsilon}$ refers to the minimum upper bound of $\varepsilon(\chi, t)$. Moreover, by magnifying p and q, $\bar{\varepsilon}$ can be made arbitrarily small.

Lemma 3.4 (Chen 2009) *Consider the FSE-RBFNN-based approximator, the estimation error can be given as*

$$\theta^T H\left(\chi, W^T G(t)\right) - \hat{\theta}^T H\left(\chi, \hat{W}^T G(t)\right)$$
$$= \tilde{\theta}^T\left(\hat{H} - \hat{H}'\hat{W}^T G(t)\right) + \hat{\theta}^T \hat{H}'\tilde{W}^T G(t) + \varrho,$$

where $\hat{H} = H(\chi, \hat{W}^T G(t))$, $\hat{H}' = [\hat{h}'_1, \hat{h}'_2, \ldots, \hat{h}'_p]^T \in \mathbb{R}^{p\times m}$ *with*

$$\hat{h}'_j = \frac{\partial h_j(\chi, d(t))}{\partial d}\Big|_{d=\hat{W}^T G(t)},\ j = 1, \ldots, p.$$

Term ϱ *is bounded by*

$$|\varrho| \le ||W||_F ||G(t)\hat{\theta}^T \hat{H}'||_F + ||\theta|| ||\hat{H}'\hat{W}^T G(t)|| + |\theta|_1$$

where $|\theta|_1 = \sum_{i=1}^{p} |\zeta_i|$.

Furthermore, define $\hat{\theta}$ and $\hat{W}$ as the estimations with the weight estimation errors $\tilde{\theta} = \theta - \hat{\theta}$ and $\tilde{W} = W - \hat{W}$, respectively. For the proof processes of Lemmas 3.3 and 3.4, please refer to the Chen (2009).

3.4 Adaptive NN Controller Design and Stability Analysis

The adaptive NN controller will be designed for nonlinear system (3.1) in this section, where unknown control direction, dead zone output, and periodic disturbances are considered at the same time. The Nussbaum-type function is applied to settle the issues of the dead zone output and unknown control direction. The FSE-RBFNN-based approximator is employed to identify the unknown nonlinearities. Moreover, the adaptive controller design consists of n steps.

Step 1: The state transformation is defined as

$$z_1 = y - y_d$$

with y_d being the desired reference signal. Choose a Lyapunov function as

$$V_1 = \frac{1}{2}z_1^2 + \frac{1}{2}\tilde{\theta}_1^T \gamma_1^{-1}\tilde{\theta}_1 + \frac{1}{2}tr\left\{\tilde{W}_1^T \delta_1^{-1}\tilde{W}_1\right\},$$

where γ_1 and δ_1 are positive definite matrices. According to Lemma 3.2, $\dot{V}_1$ is computed as

$$\dot{V}_1 = z_1 \left[\eta_1 a_1 x_2 + \eta_1 f_1(\bar{x}_1, d_1(t)) - \dot{y}_d\right] \\ -\tilde{\theta}_1^T \gamma_1^{-1} \dot{\hat{\theta}}_1 - tr\left\{\tilde{W}_1^T \delta_1^{-1} \dot{\hat{W}}_1\right\}.$$

It should be noted that η_1 is varying. Because of this factor, it makes the adaptive neural controller design and stability analysis more intractability. For addressing this trouble, we introduce the Nussbaum-type function $N(\rho_1)$. Define the first virtual controller as

$$\alpha_1 = N(\rho_1)\bar{\alpha}_1. \tag{3.8}$$

With the above mentioned, it can be obtained that

$$\dot{V}_1 = z_1 \left[\eta_1 a_1 z_2 + \eta_1 a_1 N(\rho_1)\bar{\alpha}_1 + \eta_1 f_1(\bar{x}_1, d_1(t)) - \dot{y}_d\right] \\ -\tilde{\theta}_1^T \gamma_1^{-1} \dot{\hat{\theta}}_1 - tr\left\{\tilde{W}_1^T \delta_1^{-1} \dot{\hat{W}}_1\right\},$$

where $z_2 = x_2 - \alpha_1$. Define $F_1 = \eta_1 f_1(\bar{x}_1, d_1(t)) - \dot{y}_d$. From the FSE-RBFNN-based approximator, we have

$$F_1 = \varepsilon_1(\chi_1, t) + \theta_1^T H_1\left(\chi_1, W_1^T G_1(t)\right),$$

where $\chi_1 = [\bar{x}_1, y_d, \dot{y}_d]^T$. Furthermore, in accordance with Lemmas 3.3 and 3.4, the following inequality holds

$$\begin{aligned} \dot{V}_1 = {} & z_1 \left[\hat{\theta}_1^T H_1\left(\chi_1, \hat{W}_1^T G_1(t)\right) + \hat{\theta}_1^T \hat{H}_1' \tilde{W}_1^T G_1(t) \right. \\ & \left. + \tilde{\theta}_1^T \left(\hat{H}_1 - \hat{H}_1' \hat{W}_1^T G_1(t)\right) + \varepsilon_1 + \varrho_1\right] \\ & + \eta_1 a_1 z_1 z_2 + z_1 \eta_1 a_1 N(\rho_1)\bar{\alpha}_1 - \tilde{\theta}_1^T \gamma_1^{-1} \dot{\hat{\theta}}_1 - tr\left\{\tilde{W}_1^T \delta_1^{-1} \dot{\hat{W}}_1\right\}. \end{aligned} \tag{3.9}$$

By utilizing the Lemma 3.4 and Young's inequality, one has

$$\begin{aligned} z_1 \varepsilon_1 \le {} & \frac{1}{2} z_1^2 + \frac{1}{2} \bar{\varepsilon}_1^2 \\ z_1 \varrho_1 \le {} & \frac{z_1^2}{2} ||G_1(t) \hat{\theta}_1^T \hat{H}_1'||_F^2 + \frac{1}{2} ||W_1||_F^2 + \frac{1}{2} |\theta_1|_1^2 \\ & + \frac{z_1^2}{2} ||\hat{H}_1' \hat{W}_1^T G_1(t)||^2 + \frac{1}{2} ||\theta_1||^2 + \frac{1}{2} z_1^2 \\ \eta_1 a_1 z_1 z_2 \le {} & \frac{1}{2} z_1^2 + \frac{1}{2} a_1^2 z_2^2 C_M^2. \end{aligned} \tag{3.10}$$

Design

$$\dot{\rho}_1 = z_1 \bar{\alpha}_1. \tag{3.11}$$

Along with the formulas (3.10) and (3.11), the inequality (3.9) can be rewritten as

$$\begin{aligned}\dot{V}_1 \leq\ & \dot{\rho}_1 + \eta_1 a_1 N(\rho_1)\dot{\rho}_1 + z_1 \Big[\hat{\theta}_1^T H_1 \left(\chi_1, \hat{W}_1^T G_1(t)\right) \\ & + \hat{\theta}_1^T \hat{H}_1' \tilde{W}_1^T G_1(t) + \tilde{\theta}_1^T \left(\hat{H}_1 - \hat{H}_1' \hat{W}_1^T G_1(t)\right) \\ & + \frac{z_1}{2}||G_1(t)\hat{\theta}_1^T \hat{H}_1'||_F^2 + \frac{z_1}{2}||\hat{H}_1' \hat{W}_1^T G_1(t)||^2 \\ & - \bar{\alpha}_1 + \frac{3}{2}z_1\Big] + \frac{1}{2}a_1^2 z_2^2 C_M^2 + \frac{1}{2}||W_1||_F^2 + \frac{1}{2}|\theta_1|_1^2 \\ & + \frac{1}{2}||\theta_1||^2 + \frac{1}{2}\bar{\varepsilon}_1^2 - \tilde{\theta}_1^T \gamma_1^{-1}\dot{\hat{\theta}}_1 - tr\left\{\tilde{W}_1^T \delta_1^{-1}\dot{\hat{W}}_1\right\}.\end{aligned}$$

Design the first virtual control unit $\bar{\alpha}_1$ as

$$\begin{aligned}\bar{\alpha}_1 =\ & \frac{z_1}{2}||\hat{H}_1' \hat{W}_1^T G_1(t)||^2 + \hat{\theta}_1^T H_1 \left(\chi_1, \hat{W}_1^T G_1(t)\right) \\ & + \frac{z_1}{2}||G_1(t)\hat{\theta}_1^T \hat{H}_1'||_F^2 + \kappa_1 z_1 + \frac{3}{2}z_1,\end{aligned} \tag{3.12}$$

where κ_1 is the positive constant to be designed. Then, we get

$$\begin{aligned}\dot{V}_1 \leq\ & \eta_1 a_1 N(\rho_1)\dot{\rho}_1 + \dot{\rho}_1 - \kappa_1 z_1^2 + \frac{1}{2}(|\theta_1|_1^2 + \bar{\varepsilon}_1^2 + ||W_1||_F^2 \\ & + a_1^2 z_2^2 C_M^2 + ||\theta_1||^2) - \tilde{\theta}_1^T \gamma_1^{-1}\dot{\hat{\theta}}_1 - tr\left\{\tilde{W}_1^T \delta_1^{-1}\dot{\hat{W}}_1\right\} \\ & + z_1\tilde{\theta}_1^T \left(\hat{H}_1 - \hat{H}_1' \hat{W}_1^T G_1(t)\right) + z_1\hat{\theta}_1^T \hat{H}_1' \tilde{W}_1^T G_1(t).\end{aligned}$$

It should be noted that the following formula holds

$$\hat{\theta}_1^T \hat{H}_1' \tilde{W}_1^T G_1(t) = tr\left\{\tilde{W}_1^T G_1(t)\hat{\theta}_1^T \hat{H}_1'\right\}.$$

Design the adaptive laws $\dot{\hat{\theta}}_1$ and $\dot{\hat{W}}_1$ as

$$\dot{\hat{\theta}}_1 = \gamma_1\left[z_1\left(\hat{H}_1 - \hat{H}_1' \hat{W}_1^T G_1(t)\right) - \vartheta_1\hat{\theta}_1\right] \tag{3.13}$$

$$\dot{\hat{W}}_1 = \delta_1\left[z_1 G_1(t)\hat{\theta}_1^T \hat{H}_1' - \iota_1\hat{W}_1\right], \tag{3.14}$$

where ϑ_1 and ι_1 are positive constants. Based on the above Eqs. (3.13) and (3.14), we have

$$\begin{aligned}\dot{V}_1 \leq\ & -\kappa_1 z_1^2 + a_1\eta_1 N(\rho_1)\dot{\rho}_1 + \dot{\rho}_1 + \frac{1}{2}a_1^2 z_2^2 C_M^2 \\ & + \frac{1}{2}||W_1||_F^2 + \frac{1}{2}||\theta_1||^2 + \frac{1}{2}|\theta_1|_1^2 + \frac{1}{2}\bar{\varepsilon}_1^2 \\ & + \vartheta_1\tilde{\theta}_1^T\hat{\theta}_1 + tr\left\{\iota_1\tilde{W}_1^T\hat{W}_1\right\}.\end{aligned} \tag{3.15}$$

From Young's inequality, it yields

$$\tilde{\theta}_1^T \hat{\theta}_1 \leq -\frac{1}{2}||\tilde{\theta}_1||^2 + \frac{1}{2}||\theta_1||^2 \tag{3.16}$$

$$tr\left\{\tilde{W}_1^T \hat{W}_1\right\} \leq -\frac{1}{2}||\tilde{W}_1||_F^2 + \frac{1}{2}||W_1||_F^2. \tag{3.17}$$

Substituting the above two inequalities (3.16) and (3.17) to (3.15), then the formula (3.15) can be rewritten as

$$\begin{aligned} \dot{V}_1 \leq & -\kappa_1 z_1^2 + a_1 \eta_1 N(\rho_1)\dot{\rho}_1 + \dot{\rho}_1 - \frac{\vartheta_1}{2}||\tilde{\theta}_1||^2 + \frac{1}{2}\bar{\varepsilon}_1^2 \\ & -\frac{\iota_1}{2}||\tilde{W}_1||_F^2 + \frac{1}{2}a_1^2 z_2^2 C_M^2 + \frac{1}{2}||W_1||_F^2 + \frac{1}{2}|\theta_1|_1^2 \\ & +\frac{1}{2}||\theta_1||^2 + \frac{\vartheta_1}{2}||\theta_1||^2 + \frac{\iota_1}{2}||W_1||_F^2. \end{aligned}$$

Step i ($i = 2, 3, \ldots, n-1$): Define

$$z_i = x_i - \alpha_{i-1}.$$

Design the Lyapunov function as follows:

$$V_i = V_{i-1} + \frac{1}{2}z_i^2 + \frac{1}{2}\tilde{\theta}_i^T \gamma_i^{-1}\tilde{\theta}_i + \frac{1}{2}tr\left\{\tilde{W}_i^T \delta_i^{-1}\tilde{W}_i\right\},$$

where γ_i and δ_i are the positive definite matrices. The derivative of V_i is obtained as

$$\begin{aligned} \dot{V}_i \leq & -\sum_{j=1}^{i-1}\kappa_j z_j^2 + \sum_{j=1}^{i-1}A_j N(\rho_j)\dot{\rho}_j + \sum_{j=1}^{i-1}\dot{\rho}_j + \sum_{j=1}^{i-1}\frac{1}{2}\bar{\varepsilon}_j^2 \\ & -\sum_{j=1}^{i-1}\frac{\iota_j}{2}||\tilde{W}_j||_F^2 - \sum_{j=1}^{i-1}\frac{\vartheta_j}{2}||\tilde{\theta}_j||^2 + \sum_{j=1}^{i-1}\frac{1}{2}|\theta_j|_1^2 \\ & +\sum_{j=1}^{i-1}\left(\frac{1}{2}+\frac{\vartheta_j}{2}\right)||\theta_j||^2 + \sum_{j=1}^{i-1}\left(\frac{1}{2}+\frac{\iota_j}{2}\right)||W_j||_F^2 \\ & +z_i\left\{\frac{1}{2}a_{i-1}^2 z_i + a_i x_{i+1} + f_i(\bar{x}_i, d_i(t)) - \frac{\partial\alpha_{i-1}}{\partial y}\right. \\ & \times[a_1\eta_1 x_2 + \eta_1 f_1(\bar{x}_1, d_1(t))] - \sum_{j=1}^{i-1}\frac{\partial\alpha_{i-1}}{\partial x_j}[a_j x_{j+1} \\ & +f_j(\bar{x}_j, d_j(t))] - \sum_{j=1}^{i-1}\frac{\partial\alpha_{i-1}}{\partial\rho_j}\dot{\rho}_j - \sum_{j=1}^{i-1}\frac{\partial\alpha_{i-1}}{\partial\hat{\theta}_j}\dot{\hat{\theta}}_j \end{aligned}$$

$$
\left. - \sum_{j=1}^{i-1} \frac{\partial \alpha_{i-1}}{\partial \hat{W}_j} \dot{\hat{W}}_j - \sum_{j=1}^{i-1} \frac{\partial \alpha_{i-1}}{\partial y_d^{(j-1)}} y_d^{(j)} \right\}
$$

$$
-\tilde{\theta}_i^T \gamma_i^{-1} \dot{\hat{\theta}}_i - tr\left\{ \tilde{W}_i^T \delta_i^{-1} \dot{\hat{W}}_i \right\}, \tag{3.18}
$$

where $A_1 = a_1 \eta_1$ and $A_i = a_i$ $(i = 2, \ldots, n)$.

Remark 3.2 Please pay attention to the term $\frac{1}{2} a_{i-1}^2 z_i$ in the above inequality (3.18). At Step $i = 3, \ldots, n-1$, the inequality (3.18) holds. However, at Step 2, the term $\frac{1}{2} a_{i-1}^2 z_i$ in the inequality (3.18) should be changed as $\frac{1}{2} a_1^2 z_2 C_M^2$. Accordingly, at Step 2, the term $\frac{1}{2} a_{i-1}^2 z_i$ in the following nonlinear function F_i should be changed as $\frac{1}{2} a_1^2 z_2 C_M^2$.

Define

$$
\begin{aligned}
F_i = & f_i(\bar{x}_i, d_i(t)) - \sum_{j=1}^{i-1} \frac{\partial \alpha_{i-1}}{\partial \rho_j} \dot{\rho}_j - \sum_{j=1}^{i-1} \frac{\partial \alpha_{i-1}}{\partial \hat{\theta}_j} \dot{\hat{\theta}}_j \\
& - \frac{\partial \alpha_{i-1}}{\partial y} \left[a_1 \eta_1 x_2 + \eta_1 f_1(\bar{x}_1, d_1(t)) \right] \\
& - \sum_{j=1}^{i-1} \frac{\partial \alpha_{i-1}}{\partial x_j} \left[f_j(\bar{x}_j, d_j(t)) + a_j x_{j+1} \right] \\
& - \sum_{j=1}^{i-1} \frac{\partial \alpha_{i-1}}{\partial \hat{W}_j} \dot{\hat{W}}_j - \sum_{j=1}^{i-1} \frac{\partial \alpha_{i-1}}{\partial y_d^{(j-1)}} y_d^{(j)} + \frac{1}{2} a_{i-1}^2 z_i.
\end{aligned}
$$

By using the FSE-RBFNN-based approximator, we have

$$
F_i = \theta_i^T H_i \left(\chi_i, W_i^T G_i(t) \right) + \varepsilon_i(\chi_i, t),
$$

where $\chi_i = [\bar{x}_i, \dot{\alpha}_{i-1}]^T$. The virtual controller is described by

$$
\alpha_i = N(\rho_i) \bar{\alpha}_i. \tag{3.19}
$$

Based on the definition of ϱ_i and Young's inequality, we get

$$
\begin{aligned}
a_i z_i z_{i+1} \leq & \frac{1}{2} z_i^2 + \frac{1}{2} a_i^2 z_{i+1}^2 \\
z_i \varepsilon_i \leq & \frac{1}{2} z_i^2 + \frac{1}{2} \bar{\varepsilon}_i^2 \\
z_i \varrho_i \leq & \frac{z_i^2}{2} ||G_i(t) \hat{\theta}_i^T \hat{H}_i'||_F^2 + \frac{1}{2} ||W_i||_F^2 + \frac{1}{2} |\theta_i|_1^2 \\
& + \frac{z_i^2}{2} ||\hat{H}_i' \hat{W}_i^T G_i(t)||^2 + \frac{1}{2} ||\theta_i||^2 + \frac{1}{2} z_i^2.
\end{aligned}
$$

Define $\dot{\rho}_i = z_i\bar{\alpha}_i$. Design the virtual control unit $\bar{\alpha}_i$ as

$$\bar{\alpha}_i = \kappa_i z_i + \frac{3}{2}z_i + \hat{\theta}_i^T H_i\left(\chi_i, \hat{W}_i^T G_i(t)\right) + \frac{z_i}{2}||G_i(t)\hat{\theta}_i^T \hat{H}_i^{'}||_F^2 + \frac{z_i}{2}||\hat{H}_i^{'}\hat{W}_i^T G_i(t)||^2, \tag{3.20}$$

where $\kappa_i > 0$ is the designed parameter. Then, based on formula (3.20), we have

$$\begin{aligned}\dot{V}_i \leq & -\sum_{j=1}^{i}\kappa_j z_j^2 + \sum_{j=1}^{i} A_j N(\rho)_j\dot{\rho}_j + \sum_{j=1}^{i}\dot{\rho}_j + \sum_{j=1}^{i}\frac{1}{2}\bar{\varepsilon}_j^2 \\ & + \sum_{j=1}^{i-1}\left(\frac{1}{2} + \frac{\vartheta_j}{2}\right)||\theta_j||^2 + \sum_{j=1}^{i}\frac{1}{2}|\theta_j|_1^2 + \frac{1}{2}||\theta_i||^2 \\ & + \sum_{j=1}^{i-1}\left(\frac{1}{2} + \frac{\iota_j}{2}\right)||W_j||_F^2 - \sum_{j=1}^{i-1}\frac{\iota_j}{2}||\tilde{W}_j||_F^2 \\ & - \sum_{j=1}^{i-1}\frac{\vartheta_j}{2}||\tilde{\theta}_j||^2 + \frac{1}{2}||W_i||_F^2 + z_i\hat{\theta}_i^T\hat{H}_i^{'}\tilde{W}_i^T G_i(t) \\ & + \frac{1}{2}a_i^2 z_{i+1}^2 + z_i\tilde{\theta}_i^T\left(\hat{H}_i - \hat{H}_i^{'}\hat{W}_i^T G_i(t)\right) \\ & - \tilde{\theta}_i^T\gamma_i^{-1}\dot{\hat{\theta}}_i - tr\left\{\tilde{W}_i^T\delta_i^{-1}\dot{\hat{W}}_i\right\}.\end{aligned}$$

Note that

$$\hat{\theta}_i^T\hat{H}_i^{'}\tilde{W}_i^T G_i(t) = tr\left\{\tilde{W}_i^T G_i(t)\hat{\theta}_i^T\hat{H}_i^{'}\right\}.$$

Design the adaptive laws $\dot{\hat{\theta}}_i$ and $\dot{\hat{W}}_i$ as

$$\dot{\hat{\theta}}_i = \gamma_i\left[z_i\left(\hat{H}_i - \hat{H}_i^{'}\hat{W}_i^T G_i(t)\right) - \vartheta_i\hat{\theta}_i\right] \tag{3.21}$$

$$\dot{\hat{W}}_i = \delta_i\left[z_i G_i(t)\hat{\theta}_i^T\hat{H}_i^{'} - \iota_i\hat{W}_i\right], \tag{3.22}$$

where ϑ_i and ι_i are the positive constants. Based on Young's inequality, we have

$$\begin{aligned}\tilde{\theta}_i^T\hat{\theta}_i \leq & -\frac{1}{2}||\tilde{\theta}_i||^2 + \frac{1}{2}||\theta_i||^2 \\ tr\left\{\tilde{W}_i^T\hat{W}_i\right\} \leq & -\frac{1}{2}||\tilde{W}_i||_F^2 + \frac{1}{2}||W_i||_F^2.\end{aligned}$$

Finally, we have

$$\dot{V}_i \leq -\sum_{j=1}^{i} \kappa_j z_j^2 + \sum_{j=1}^{i} A_j N(\rho_j)\dot{\rho}_j + \sum_{j=1}^{i} \dot{\rho}_j + \sum_{j=1}^{i} \frac{\bar{\varepsilon}_j^2}{2}$$
$$+\sum_{j=1}^{i} \left(\frac{1}{2} + \frac{\vartheta_j}{2}\right) ||\theta_j||^2 - \sum_{j=1}^{i} \frac{\iota_j}{2} ||\tilde{W}_j||_F^2$$
$$+\sum_{j=1}^{i} \left(\frac{1}{2} + \frac{\iota_j}{2}\right) ||W_j||_F^2 - \sum_{j=1}^{i} \frac{\vartheta_j}{2} ||\tilde{\theta}_j||^2$$
$$+\sum_{j=1}^{i} \frac{1}{2} |\theta_j|_1^2 + \frac{1}{2} a_i^2 z_{i+1}^2.$$

Step n: Define

$$z_n = x_n - \alpha_{n-1}.$$

The Lyapunov function is chosen as

$$V_n = \frac{1}{2} z_n^2 + \frac{1}{2} tr\left\{\tilde{W}_n^T \delta_n^{-1} \tilde{W}_n\right\} + \frac{1}{2} \tilde{\theta}_n^T \gamma_n^{-1} \tilde{\theta}_n + V_{n-1},$$

where δ_n and γ_n are the positive definite matrices. Define

$$u = N(\rho_n)\bar{\alpha}_n \tag{3.23}$$
$$\dot{\rho}_n = z_n \bar{\alpha}_n.$$

The virtual control unit $\bar{\alpha}_n$ is designed as

$$\bar{\alpha}_n = \hat{\theta}_n^T H_n\left(\chi_n, \hat{W}_n^T G_n(t)\right) + \frac{z_n}{2} ||\hat{H}_n' \hat{W}_n^T G_n(t)||^2$$
$$+\kappa_n z_n + z_n + \frac{z_n}{2} ||G_n(t) \hat{\theta}_n^T \hat{H}_n'||_F^2, \tag{3.24}$$

where κ_n is a positive parameter. For the positive constants ϑ_n and ι_n, design the adaptive laws $\dot{\hat{\theta}}_n$ and $\dot{\hat{W}}_n$ as

$$\dot{\hat{\theta}}_n = \gamma_n \left[z_n\left(\hat{H}_n - \hat{H}_n' \hat{W}_n^T G_n(t)\right) - \vartheta_n \hat{\theta}_n\right] \tag{3.25}$$
$$\dot{\hat{W}}_n = \delta_n \left[z_n G_n(t) \hat{\theta}_n^T \hat{H}_n' - \iota_n \hat{W}_n\right]. \tag{3.26}$$

Therefore, we get

$$\dot{V}_n \leq -\sum_{i=1}^{n} \kappa_i z_i^2 + \sum_{i=1}^{n} A_i N(\rho_i)\dot{\rho}_i + \sum_{i=1}^{n} \dot{\rho}_i$$

$$-\sum_{i=1}^{n}\frac{\iota_i}{2}||\tilde{W}_i||_F^2-\sum_{i=1}^{n}\frac{\vartheta_i}{2}||\tilde{\theta}_i||^2+D_0,$$

where $D_0=\sum_{i=1}^{n}\left(\frac{1}{2}+\frac{\vartheta_i}{2}\right)||\theta_i||^2+\sum_{i=1}^{n}\frac{1}{2}|\theta_i|_1^2+\sum_{i=1}^{n}\frac{1}{2}\bar{\varepsilon}_i^2+\sum_{i=1}^{n}\left(\frac{1}{2}+\frac{\iota_i}{2}\right)||W_i||_F^2$.

Finally, with the above backstepping processes, the following theorem is given.

Theorem 3.1 *Consider the strict-feedback nonlinear system (3.1) in the presence of periodic disturbances, dead zone output, and unknown control direction. Under Assumption 3.1, the designed control laws (3.8), (3.12), (3.19), (3.20), (3.23), (3.24) and adaptive laws (3.13), (3.14), (3.21), (3.22), (3.25), (3.26), then it is proved that all the closed-loop signals are bounded and the tracking error converges to zero in a small neighborhood.*

Proof The following expression holds

$$\dot{V}_n\leq -DV_n+\sum_{i=1}^{n}A_iN(\rho_i)\dot{\rho}_i+\sum_{i=1}^{n}\dot{\rho}_i+D_0, \tag{3.27}$$

where $D=\min\{2\kappa_i,\ \vartheta_i/\lambda_{\max}(\gamma_i^{-1}),\ \iota_i/\lambda_{\max}(\delta_i^{-1}),\ i=1,\ldots,n\}$. Multiplying the formula (3.27) by e^{Dt} on both sides leads to

$$e^{Dt}\dot{V}_n(t)\leq\ -DV_ne^{Dt}+e^{Dt}\sum_{i=1}^{n}A_iN(\rho_i)\dot{\rho}_i+e^{Dt}\sum_{i=1}^{n}\dot{\rho}_i+D_0e^{Dt}$$

and

$$\frac{d(V_n(t)e^{Dt})}{dt}\leq\ e^{Dt}\sum_{i=1}^{n}A_iN(\rho_i)\dot{\rho}_i+D_0e^{Dt}+e^{Dt}\sum_{i=1}^{n}\dot{\rho}_i. \tag{3.28}$$

Integrating the formula (3.28) over $[0,t]$ leads to

$$V_n(t)\leq\ V_n(0)e^{-Dt}+e^{-Dt}\sum_{i=1}^{n}\int_0^t A_iN(\rho_i)\dot{\rho}_ie^{D\tau}d\tau$$
$$+\frac{D_0}{D}+e^{-Dt}\sum_{i=1}^{n}\int_0^t\dot{\rho}_ie^{D\tau}d\tau.$$

Applying Lemma 3.1, the boundedness of $V_i(t)$ and $\rho_i(t)$ is guaranteed with $i=1,\ldots,n$. Furthermore, it can be seen that

$$V_n(t)\leq\frac{D_0}{D}+V_n(0)e^{-Dt}+B_n$$

where B_n is the upper bound of $e^{-Dt}\sum_{i=1}^{n}\int_0^t A_i N(\rho_i)\dot{\rho}_i e^{D\tau}d\tau + e^{-Dt}\sum_{i=1}^{n}\int_0^t \dot{\rho}_i e^{D\tau}d\tau$. Furthermore, one has

$$\lim_{t\to\infty}|z_1(t)| \le \sqrt{2\left(\frac{D_0}{D}+B_n\right)}.$$

By designing adequate parameters, we can make the tracking error z_1 arbitrarily small. The proof is completed. □

Remark 3.3 In Ma and Ma (2021), only the unknown coefficient problem is considered, which can not simultaneously deal with the unknown control direction and output dead zone issues. As a result, in order to simplify calculation, these two cases are handled by a Nussbaum framework. Besides, the investigated model is general since when unknown coefficients are set to one, it will be changed as the common form (Wang et al. 2020; Du et al. 2021).

Remark 3.4 In the traditional methods (Wang et al. 2015; Wang and Chen 2020; Ma et al. 2019; Cao et al. 2019; Dong et al. 2021; Du et al. 2021), the external disturbances $d_i(t)$ $(i = 1, 2, \ldots, n)$, and nonlinear functions $f_i(\bar{x}_i)$ $(i = 1, 2, \ldots, n)$ are processed separately, i.e., $d_i(t)$ $(i = 1, 2, \ldots, n)$ are addressed by an upper bound and $f_i(\bar{x}_i)$ $(i = 1, 2, \ldots, n)$ are solved by FLS/NN. However, the above schemes can not be applied to this work due to that nonlinear functions include both the state variables and time-varying periodic disturbances. Hence, this chapter considers to use FSE and RBFNN techniques to tackle the corresponding problem.

3.5 Simulation Results

In this section, two examples are given to demonstrate the effectiveness of the proposed approach.

Example 3.1 Consider the nonlinear system as

$$\begin{aligned}\dot{x}_1 &= a_1x_2 + 0.8\cos(2x_1d_1(t))\\ \dot{x}_2 &= a_2u + \sin(x_1^2x_2d_2(t))e^{(-x_1^2d_2(t))}\\ y &= \psi(x_1),\end{aligned}$$

where the time-varying disturbances are $d_1(t) = |\sin(\frac{1}{2}t)|$ and $d_2(t) = |\cos(t)|$ with known periods $T_1 = 2\pi$ and $T_2 = \pi$, respectively. Moreover, the reference signal is set as $y_d = 0.7\sin(t)$. The dead zone parameters are chosen as $\bar{m}_r = 0.71$, $\bar{m}_l = 0.68$, $c_r = 0.1$, $c_l = -0.1$. The parameters of the adaptive laws and designed controllers are chosen as $\kappa_1 = 15$, $\kappa_2 = 10$, $\vartheta_1 = 0.5$, $\vartheta_2 = 0.01$, $\iota_1 = 0.01$, $\iota_2 = 0.001$, $\gamma_1 = 0.1I$, $\gamma_2 = 0.5I$, $\delta_1 = 0.001I$, $\delta_2 = 2.2I$, $a_1 = 1.20$, $a_2 = 8$. $G_1(t) = [1, \sqrt{2}\sin(t), \sqrt{2}\cos(t), \sqrt{2}\sin(2t), \sqrt{2}\cos(2t)]^T$, $G_2(t) =$

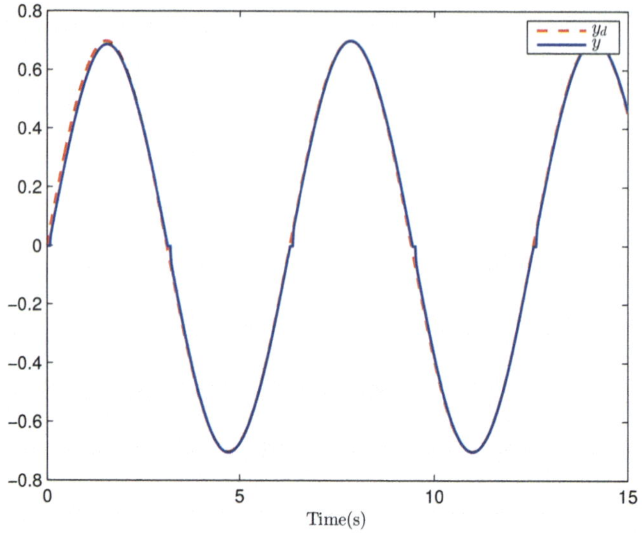

Fig. 3.1 The trajectories of reference signal y_d and system output y

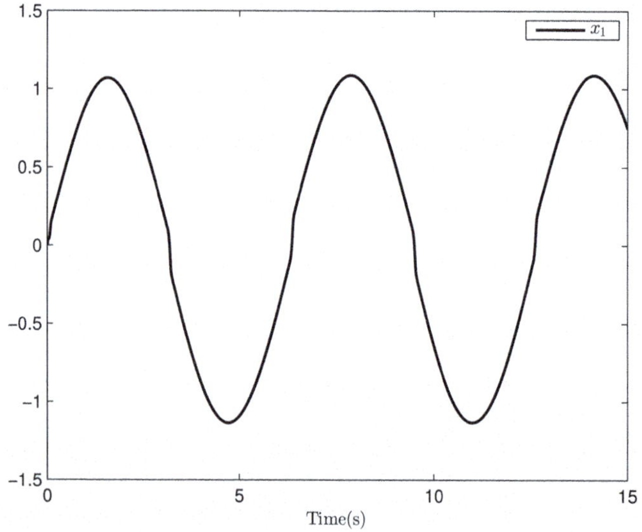

Fig. 3.2 The trajectory of system state x_1

$[1, \sqrt{2}\sin(2t), \sqrt{2}\cos(2t), \sqrt{2}\sin(4t), \sqrt{2}\cos(4t)]^T$. The simulation results are shown from Figs. 3.1, 3.2, 3.3, 3.4, and 3.5.

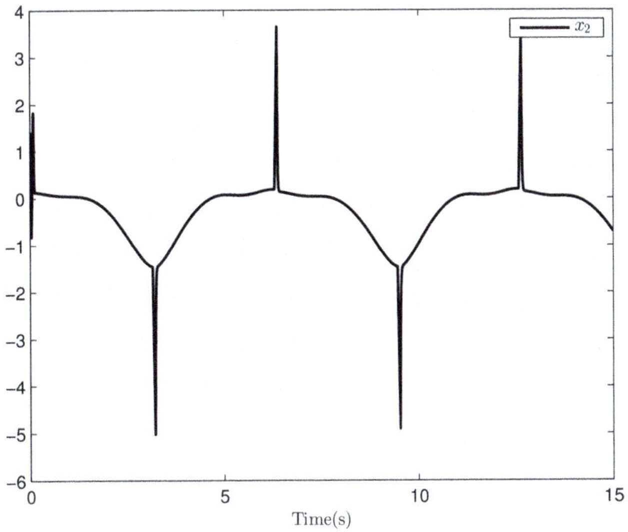

Fig. 3.3 The trajectory of system state x_2

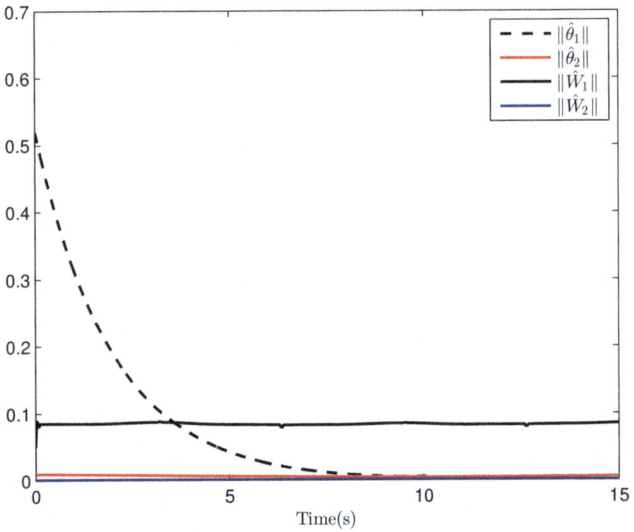

Fig. 3.4 The trajectories of adaptive parameters

Example 3.2 Consider the van der Pol oscillator as

$$\begin{aligned}\dot{x}_1 &= a_1 x_2 - \frac{1}{3}x_1^3 + x_1 + \breve{p} + \breve{d}(t)\\ \dot{x}_2 &= a_2 u + 0.1(x_1 + \breve{a} - \breve{b}x_2)\\ y &= \psi(x_1),\end{aligned}$$

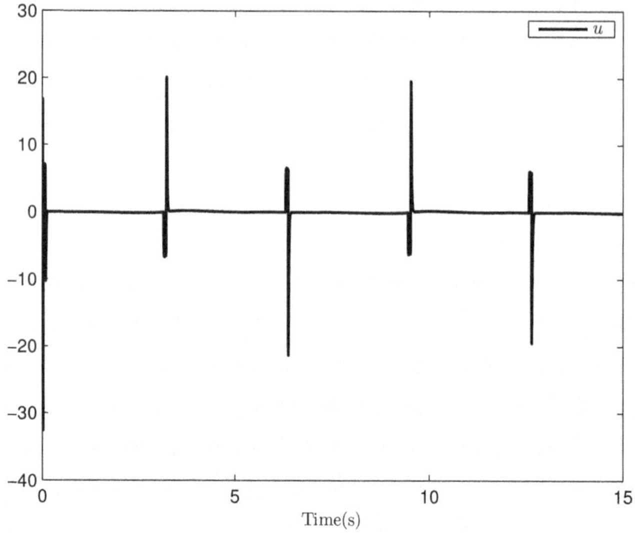

Fig. 3.5 The trajectory of control input u

where $\breve{d}(t) = q\cos(\omega t)$ denotes the periodic exciting signal. As given in Chen et al. (2010), the van der Pol oscillator parameters are chosen as $a_1 = -1, a_2 = 1, \breve{p} = 0$, $q = 0.74$, $\omega = 1$, $\breve{a} = 0.7$, and $\breve{b} = 0.8$. The dead zone parameters are chosen as $\bar{m}_r = \bar{m}_l = 0.1, c_r = 0.2, c_l = -0.2$. The reference signal and the design parameters $\vartheta_1, \vartheta_2, \iota_1, \gamma_1, \gamma_2, \delta_1, \delta_2$ are same as those given in Example 3.1. Moreover, $\kappa_1 = \kappa_2 = 60$ and $\iota_2 = 10$. For Example 3.2, the simulation results are shown from Figs. 3.6, 3.7, 3.8, 3.9 and 3.10.

Figures 3.1 and 3.6 display the trajectories of system output y and the desired reference signal y_d. Figures 3.2 and 3.3 reveal the responses of states x_1 and x_2 for Example 3.1. The trajectories of states x_1 and x_2 are shown in Figs. 3.7 and 3.8 for Example 3.2. Moreover, Figs. 3.4 and 3.9 show the responses of adaptive parameters $\|\hat{\theta}_1\|$, $\|\hat{\theta}_2\|$, $\|\hat{W}_1\|$ and $\|\hat{W}_2\|$. Figures 3.5 and 3.10 represent control input u. For two different simulation Examples 3.1 and 3.2, we try to choose the same design parameters to show that our proposed control method ensures that the system output y can track the same reference signal y_d. Based on the Examples 3.1 and 3.2, it should be highlighted that the proposed Nussbaum gain adaptive control method is able to guarantee the stability of the system (3.1) with periodic disturbances and dead zone output.

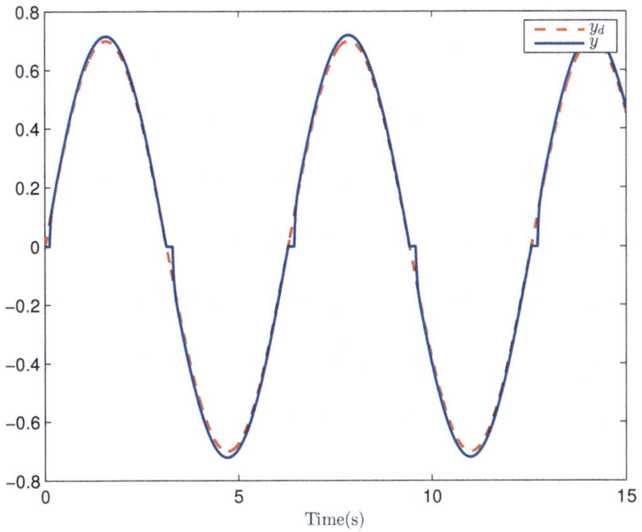

Fig. 3.6 The trajectories of reference signal y_d and system output y

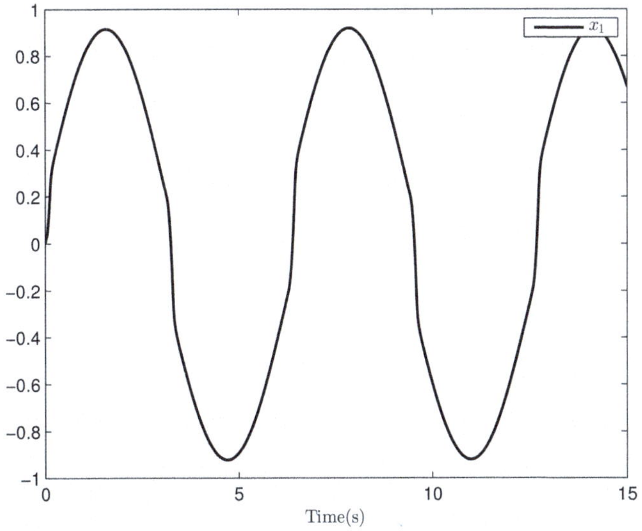

Fig. 3.7 The trajectory of system state x_1

3.6 Conclusion

An adaptive NN control strategy has been proposed for a nonlinear strict-feedback system subject to periodic disturbances, dead zone output, and unknown control direction in this chapter. Through utilizing the FSE-RBFNN-based approximator,

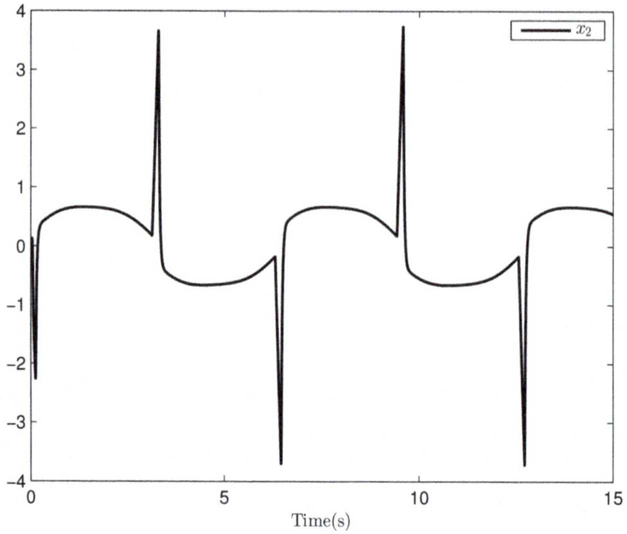

Fig. 3.8 The trajectory of system state x_2

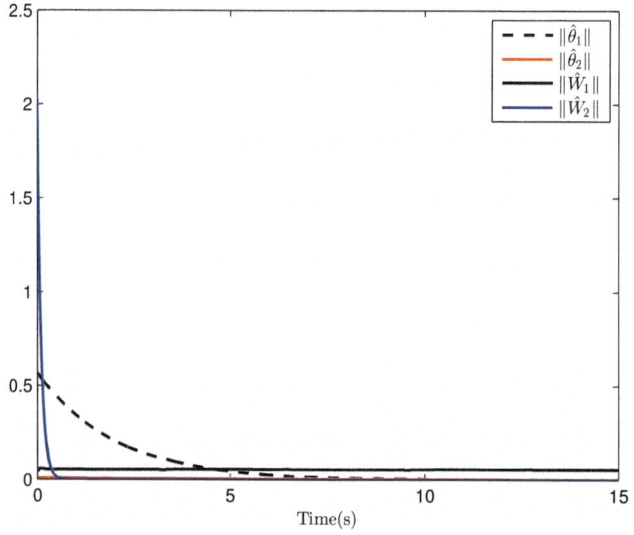

Fig. 3.9 The trajectories of adaptive parameters

the unknown nonlinear functions have been approximated effectively. With the help of the Nussbaum-type function, dead zone output and unknown control direction problems have been handled successfully, which paves a way for developing the adaptive NN control strategy under unknown nonlinearities. Moreover, it has been

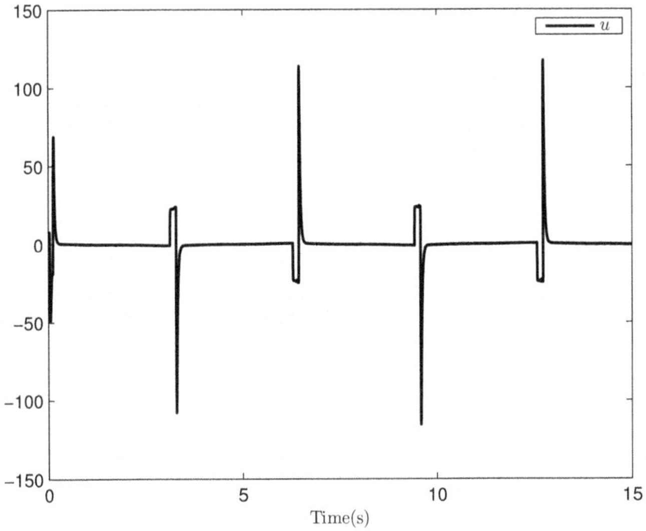

Fig. 3.10 The trajectory of control input u

proved that the system output signal can track the desired reference signal and all signals of the resulting closed-loop system are bounded. At last, simulation results have been presented to prove the effectiveness of the designed control strategy.

References

Bai W, Zhou Q, Li T, Li H (2020) Adaptive reinforcement learning neural network control for uncertain nonlinear system with input saturation. IEEE Trans Cybernet 50(8):3433–3443

Cao L, Li H, Wang N, Zhou Q (2019) Observer-based event-triggered adaptive decentralized fuzzy control for nonlinear large-scale systems. IEEE Trans Fuzzy Syst 27(6):1201–1214

Chen W (2009) Adaptive backstepping dynamic surface control for systems with periodic disturbances using neural networks. IET Control Theory Appl 3(10):1383–1394

Chen W, Jiao L, Li R, Li J (2010) Adaptive backstepping fuzzy control for nonlinearly parameterized systems with periodic disturbances. IEEE Trans Fuzzy Syst 18(4):674–685

Chen B, Liu X, Lin C (2017) Observer and adaptive fuzzy control design for nonlinear strict-feedback systems with unknown virtual control coefficients. IEEE Trans Fuzzy Syst 26(3):1732–1743

Dong G, Li H, Ma H, Lu R (2021) Finite-time consensus tracking neural network FTC of multi-agent systems. IEEE Trans Neural Netw Learn Syst 32(2):653–662

Du P, Pan Y, Li H, Lam HK (2021) Nonsingular finite-time event-triggered fuzzy control for large-scale nonlinear systems. IEEE Trans Fuzzy Syst 29(8):2088–2099

Ge SS, Wang J (2003) Robust adaptive tracking for time-varying uncertain nonlinear systems with unknown control coefficients. IEEE Trans Autom Control 48(8):1463–1469

Ge SS, Wang C, Lee TH (2000) Adaptive backstepping control of a class of chaotic systems. Int J Bifurcat Chaos 10(05):1149–1156

He W, Dong Y (2018) Adaptive fuzzy neural network control for a constrained robot using impedance learning. IEEE Trans Neural Netw Learn Syst 29(4):1174–1186

Hu X, Wang J (2006) Solving pseudomonotone variational inequalities and pseudoconvex optimization problems using the projection neural network. IEEE Trans Neural Netw 17(6):1487–1499
Hu Z, Deng F, Wu Z (2021) Synchronization of stochastic complex dynamical networks subject to consecutive packet dropouts. IEEE Trans Cybernet 51(7):3779–3788
Hu Z, Deng F, Shi P, Lim CC (2021) A new approach to characterize successive packet losses in stochastic networked systems. IEEE Trans Syst Man Cybernet Syst 51(1):552–560
Huang YB, He Y, An J, Wu M (2021) Polynomial-type lyapunov-krasovskii functional and jacobi-bessel inequality: further results on stability analysis of time-delay systems. IEEE Trans Autom Control 66(6):2905–2912
Kanellakopoulos I, Kokotovic PV, Morse AS (1991) Systematic design of adaptive controllers for feedback linearizable systems. Am Control Conf 649–654
Krstić M, Kanellakopoulos I, Kokotović P (1992) Adaptive nonlinear control without overparametrization. Syst Control Lett 19(3):177–185
Li D, Chen CLP, Liu YJ, Tong S (2019) Neural network controller design for a class of nonlinear delayed systems with time-varying full-state constraints. IEEE Trans Neural Netw Learn Syst 30(9):2625–2636
Li H, Wu Y, Chen M (2021) Adaptive fault-tolerant tracking control for discrete-time multiagent systems via reinforcement learning algorithm. IEEE Trans Cybernet 51(3):1163–1174
Liang H, Zhang Y, Huang T, Ma H (2020) Prescribed performance cooperative control for multiagent systems with input quantization. IEEE Trans Cybernet 50(5):1810–1819
Lin G, Li H, Ma H, Yao D, Lu R (2022) Human-in-the-loop consensus control for nonlinear multi-agent systems with actuator faults. IEEE/CAA J Automatica Sinica 9(1):111–122
Liu W, Li P (2019) Disturbance observer-based fault-tolerant adaptive control for nonlinearly parameterized systems. IEEE Trans Indus Electron 66(11):8681–8691
Liu YJ, Tong S (2017) Barrier Lyapunov functions for Nussbaum gain adaptive control of full state constrained nonlinear systems. Automatica 76:143–152
Liu Q, Wang J (2015) A second-order multi-agent network for bound-constrained distributed optimization. IEEE Trans Autom Control 60(12):3310–3315
Liu Z, Lai G, Zhang Y, Chen CLP (2015) Adaptive fuzzy tracking control of nonlinear time-delay systems with dead-zone output mechanism based on a novel smooth model. IEEE Trans Fuzzy Syst 23(6):1998–2011
Liu W, Lim CC, Shi P, Xu S (2017) Observer-based tracking control for MIMO pure-feedback nonlinear systems with time-delay and input quantisation. Int J Control 90(11):2433–2448
Liu Y, Liu X, Jing Y, Zhang Z (2018) A novel finite-time adaptive fuzzy tracking control scheme for nonstrict feedback systems. IEEE Trans Fuzzy Syst 27(4):646–658
Liu L, Gao T, Liu YJ, Tong S (2020) Time-varying asymmetrical blfs based adaptive finite-time neural control of nonlinear systems with full state constraints. IEEE/CAA J Automatica Sinica 7(5):1335–1343
Liu Q, Leng J, Yan D, Zhang D, Wei L, Yu A, Zhao R, Zhang H, Chen X (2021) Digital twin-based designing of the configuration, motion, control, and optimization model of a flow-type smart manufacturing system. J Manuf Syst 58:52–64
Liu Y, Yao D, Li H, Lu R (2022) Distributed cooperative compound tracking control for a platoon of vehicles with adaptive NN. IEEE Trans Cybernet 52(7):7039–7048
Ma Z, Ma H (2021) Improved adaptive fuzzy output-feedback dynamic surface control of nonlinear systems with unknown dead-zone output. IEEE Trans Fuzzy Syst 29(8):2122–2131
Ma H, Liang H, Zhou Q, Ahn CK (2019) Adaptive dynamic surface control design for uncertain nonlinear strict-feedback systems with unknown control direction and disturbances. IEEE Trans Syst Man Cybernet Syst 49(3):506–515
Ma H, Li H, Lu R, Huang T (2020) Adaptive event-triggered control for a class of nonlinear systems with periodic disturbances. Sci China Inf Sci 63(5):150212
Muramatsu H, Katsura S (2018) An adaptive periodic-disturbance observer for periodic-disturbance suppression. IEEE Trans Indus Inf 14(10):4446–4456

Nussbaum RD (1983) Some remarks on a conjecture in parameter adaptive control. Syst Control Lett 3(5):243–246

Park J, Sandberg IW (1991) Universal approximation using radial-basis-function networks. Neural Comput 3(2):246–257

Roy TK, Mahmud MA, Oo AM (2018) Robust adaptive backstepping excitation controller design for higher-order models of synchronous generators in multimachine power systems. IEEE Trans Power Syst 34(1):40–51

Sokolov Y, Kozma R, Werbos LD, Werbos PJ (2015) Complete stability analysis of a heuristic approximate dynamic programming control design. Automatica 59:9–18

Song YD, Huang X, Wen C (2017) Robust adaptive fault-tolerant PID control of MIMO nonlinear systems with unknown control direction. IEEE Trans Indus Electron 64(6):4876–4884

Tong S, Min X, Li Y (2020) Observer-based adaptive fuzzy tracking control for strict-feedback nonlinear systems with unknown control gain functions. IEEE Trans Cybernet 50(9):3903–3913

Wang L, Chen CLP (2020) Adaptive fuzzy dynamic surface control of nonlinear constrained systems with unknown virtual control coefficients. IEEE Trans Fuzzy Syst 28(8):1737–1747

Wang F, Liu Z, Lai G (2015) Fuzzy adaptive control of nonlinear uncertain plants with unknown dead zone output. Fuzzy Sets Syst 263:27–48

Wang W, Wang D, Peng Z, Li T (2016) Prescribed performance consensus of uncertain nonlinear strict-feedback systems with unknown control directions. IEEE Trans Syst Man Cybernet Syst 46(9):1279–1286

Wang L, Chen CLP, Li H (2020) Event-triggered adaptive control of saturated nonlinear systems with time-varying partial state constraints. IEEE Trans Cybernet 50(4):1485–1497

Wang J, Shi L, Guan X (2020) Semi-global leaderless consensus of linear multi-agent systems with actuator and communication constraints. J Syst Sci Complexity 33(4):882–902

Wu C, Wu L, Liu J, Jiang ZP (2020) Active defense-based resilient sliding mode control under denial-of-service attacks. IEEE Trans Inf Forensics Secur 15:237–249

Wu Y, Liang H, Zhang Y, Ahn CK (2021) Cooperative adaptive dynamic surface control for a class of high-order stochastic nonlinear multiagent systems. IEEE Trans Cybernet 51(11):5214–5224

Xia J, Zhang J, Sun W, Zhang B, Wang Z (2018) Finite-time adaptive fuzzy control for nonlinear systems with full state constraints. IEEE Trans Syst Man Cybernet Syst 49(7):1541–1548

Zeng Z, Wang J (2006) Improved conditions for global exponential stability of recurrent neural networks with time-varying delays. IEEE Trans Neural Netw 17(3):623–635

Zeng Z, Wang J (2008) Design and analysis of high-capacity associative memories based on a class of discrete-time recurrent neural networks. IEEE Trans Syst Man Cybernet Part B (Cybernetics) 38(6):1525–1536

Zhang G, Zhang X (2016) Concise adaptive fuzzy control of nonlinearly parameterized and periodically time-varying systems via small gain theory. Int J Control Autom Syst 14(4):893–905

Zhang CK, He Y, Jiang L, Wu M, Zeng HB (2016) Delay-variation-dependent stability of delayed discrete-time systems. IEEE Trans Autom Control 61(9):2663–2669

Zhang CK, Long F, He Y, Yao W, Wu M (2020) A relaxed quadratic function negative-determination lemma and its application to time-delay systems. Automatica 113:108764

Zhao Q, Sun J, Bai Y (2020) Dynamic event-triggered control for nonlinear systems: a small-gain approach. J Syst Sci Complexity 33(4):930–943

Zhou Q, Zhao S, Li H, Lu R, Wu C (2019) Adaptive neural network tracking control for robotic manipulators with dead zone. IEEE Trans Neural Netw Learn Syst 30(12):3611–3620

Zhu S, Wang M, Liu Y (2017) Adaptive decentralised control of fluid networks with random disturbances. IET Control Theory Appl 11(18):3321–3328

Zhu S, Liu D, Yang C, Fu J (2020) Synchronization of memristive complex-valued neural networks with time delays via pinning control method. IEEE Trans Cybernet 50(8):3806–3815

Zuo R, Dong X, Liu Y, Liu Z, Zhang W (2019) Adaptive neural control for MIMO pure-feedback nonlinear systems with periodic disturbances. IEEE Trans Neural Netw Learn Syst 30(6):1756–1767

Chapter 4
Adaptive Dynamic Event-Triggered Control Using Disturbance Observer

Abstract Focusing on one class of nonlinear systems with input saturation, the adaptive fuzzy dynamic event-triggered control problem is successfully solved. A dynamic event-triggered mechanism is adopted to save the communication burden of the system, and is integrated into the controller design process based on the backstepping technique. For handling the asymmetric input saturation, an auxiliary system with the same order as the considered system is constructed. Analysis problems caused by unknown continuous functions are efficiently tackled by applying the approximation theory of fuzzy logic systems. It is proven that all the signals of the resulting closed-loop system are bounded via Lyapunov stability analysis. Both numerical and practical simulations are presented to demonstrate the effectiveness of the theoretical results.

4.1 Introduction

Over the last few decades, the adaptive intelligent control theory and application of nonlinear systems have developed rapidly (Lei et al. 2022; He et al. 2020; Chen et al. 2020; Yang et al. 2018; Pan et al. 2022). For example, for a fully actuated autonomous underwater vehicle (Cui et al. 2017), an adaptive tracking control method was proposed based on discrete-time dynamics model. As one of the frontier topics of adaptive control theory and application, the backstepping technique (Wu et al. 2019) has been widely utilized to solve relatively complex nonlinear system control problems. The work in Ge et al. (2000) successfully solved the adaptive backstepping control problem for a class of chaotic systems, which adopted a systematic design method for establishing associated Lyapunov functions and feedback control laws. By combining adaptive control and neural networks (NNs) or fuzzy logic systems (FLSs), adaptive neural control and adaptive fuzzy control methods have attracted a lot of attention, see, for example, Pan et al. (2022), Wang et al. (2022), Lin et al. (2022), Li et al. (2021), Liu et al. (2022), and references therein. By incorporating a dynamic signal and the property of the NNs into the adaptive backstepping recursive design, an adaptive control law for a class of nonaffine nonlinear multi-agent systems was designed in Liang et al. (2021). To remove the restrictive assumptions that the

H. Ren et al., *Intelligent Control of Nonlinear Systems and Nonlinear Multi-Agent Systems with Complex Constraints*, Intelligent Control and Learning Systems 16,
https://doi.org/10.1007/978-981-96-0320-6_4

virtual control gains must be constants, the adaptive fuzzy output-feedback controller was developed in Tong et al. (2020) for uncertain strict-feedback nonlinear systems by constructing logarithm Lyapunov functions.

The event-triggered control scheme is an effective means to reduce the communication burden while ensuring system performance (Heemels et al. 2013; Hu et al. 2021; Wang et al. 2021; Zhang et al. 2019; Chen et al. 2022; Li et al. 2020). The design of event-triggered mechanism is the main difficult problem affecting the event-triggered control performance, especially for the nonstrict-feedback nonlinear system with external disturbance (Hu et al. 2021; Liu et al. 2021). Starting from the control and estimation strategies and system dynamics, the review paper (Chen et al. 2020) gave a unified formulation of the existing triggering technologies so that the triggering mechanisms can be better understood. To reduce unnecessary data communications in Ding et al. (2017), a state-independent threshold event-triggered mechanism was adopted to update control signals of multi-agent systems against cyber-attacks and lossy sensors. Xu and Wu (2021) investigated an observer-based distributed adaptive fault-tolerant synchronization problem, in which the parameters of triggering threshold online were adjusted by state-dependent and time-dependent thresholds. Based on a relative threshold event-triggered mechanism, a bipartite adaptive tracking control strategy was formulated in Liang et al. (2021) for stochastic nonlinear systems using FLSs. In order to control the constrained quarter active vehicle suspension systems within time-variant constraints, the authors designed a novel event-triggered fixed-time convergence controller in Jia et al. (2022). A dynamic event-triggering mechanism was designed to reduce the consumption of communication bandwidth in Wang and Chen (2021), in which an improved NNs that considered reconstructed error has been used for identifying the unknown nonlinear terms.

Many control devices have the characteristics of uncertainty and nonlinearity, which widely exist in practical engineering (Zhang et al. 2022). In addition, in view of the inherent physical limitations of actuator, the amplitude of the actual control input signal to the control system is always limited. As a common constraint for control inputs with nonsmoothness and nonlinearity, the study of saturation is of great significance in controller design. Recently, extensive research has been conducted on various classes of systems with input saturation (Shen and Lam 2022; Gao et al. 2022; Guo et al. 2019; Huang and Cao 2019). A hierarchical sliding-mode controller was designed in Zhao et al. (2017), in which the part beyond input saturation was compensated via an appropriate Lyapunov functional in controller design process. An effective impedance controller was designed to reduce the saturation effect in He et al. (2016), and the developed learning approach did not require the knowledge of robotic dynamics. A finite-time distributed formation control strategy was investigated in Huang et al. (2021) for unmanned surface vessels exposed to external disturbances, model uncertainties and input saturation constraints. In Hu et al. (2021), an active fault-tolerant control system was proposed for spacecrafts subject to model uncertainty, external disturbance and input saturation simultaneously. Through the authors' extensive and in-depth research, the dynamic event-triggered control problem has not been well solved for a class of nonlinear nonstrict-feedback systems with unknown disturbance and input saturation.

Inspired by the discussion above and considering a dynamic event-triggered mechanism, a disturbance observer-based fuzzy adaptive control method is designed for the system with input saturation. The main contributions of this chapter are summarized as follows:

1. For reducing some unnecessary transmissions and saving communication resources, a dynamic event-triggered mechanism is designed in this chapter. Compared with the work in Ma et al. (2020), in which the coefficient of the control input is a fixed value with an open interval from 0 to 1, the threshold parameter in the event-triggering condition is adjusted dynamically based on a dynamic rule in this chapter.
2. In order to handle the unknown external disturbance, the nonlinear disturbance observer in this chapter is designed with auxiliary variables. The designed nonlinear disturbance observer can ensure that the estimation error of disturbance converges to a bounded compact set.
3. By constructing an auxiliary system and using Lyapunov stability method, the stability of the closed-loop system with input saturation is guaranteed, and it shows that all signals of closed-loop system are semi-global uniformly ultimately bounded, and good control effect is obtained.

4.2 Problem Description and Preliminary Knowledge

4.2.1 Problem Statement

Consider a class of nonstrict-feedback nonlinear system with input saturation and external disturbance:

$$\begin{aligned} \dot{x}_i &= x_{i+1} + f_i(x) \\ \dot{x}_n &= u(v(t)) + f_n(x) + d(t) \\ y &= x_1, \quad 1 \le i \le n-1, \end{aligned} \tag{4.1}$$

where $x = [x_1, x_2, \ldots, x_n]^{\mathrm{T}} \in \mathbb{R}^n$ stands for the system state, $f_i(x)$, $i = 1, 2, \ldots, n$, is the unknown smooth nonlinear function, $y \in \mathbb{R}$ denotes the system output, $d(t) \in \mathbb{R}$ means the external disturbance, $v(t)$ is the actuator input, and $u(v(t))$ represents the system input with saturation nonlinearity, which is expressed as

$$u(v(t)) = \begin{cases} \vartheta_{\max}, & v(t) > \vartheta_{\max} \\ v(t), & \vartheta_{\min} \le v(t) \le \vartheta_{\max} \\ \vartheta_{\min}, & v(t) < \vartheta_{\min} \end{cases}$$

where $\vartheta_{\max} > 0$ and $\vartheta_{\min} < 0$ are the maximum and minimum of the actuator output, respectively.

The following assumptions are introduced for facilitating the disturbance observer-based controller design of nonlinear system (4.1).

Assumption 4.1 (Wang and Chen 2021) The control system is bounded-input bounded-state.

Assumption 4.2 (Chen et al. 2015, 2021) The external disturbance satisfies $|d(t)| \leq \bar{d}_0$ and $|\dot{d}(t)| \leq \bar{d}_1$ with positive constants $\bar{d}_0$ and $\bar{d}_1$.

4.2.2 Dynamic Event-Triggering Algorithm

In order to reduce the frequency of controller update and save energy or communication costs, for $k \in Z^+$, define the following dynamic event-triggered mechanism:

$$\begin{aligned} \upsilon(t) &= \varpi(t_k),\ \forall t \in [t_k, t_{k+1}) \\ \dot{\varsigma}(t) &= -a\varsigma^{\iota}(t) \\ t_{k+1} &= \inf\Big\{t > t_k \mid |e(t)| > \varsigma(t)|\upsilon(t)| + \sigma\Big\}, \end{aligned} \tag{4.2}$$

where $\iota \geq 2$ is a design parameter and σ is a positive constant. For any given initial value $\varsigma(0) \in (0, 1)$, we can get $0 < \varsigma(t) < 1$ for $\forall t \geq 0$. Moreover, $\varpi(t)$ represents the current actuator input, and $e(t) = \varpi(t) - \upsilon(t)$ is the measurement error between the sampled actuator input and $\varpi(t)$. The control signal holds as $\varpi(t_k)$ during the time $t \in [t_k, t_{k+1})$. The time will be marked as t_{k+1}, and the actuator input $\upsilon(t_{k+1})$ will act directly on the system when the dynamic event-triggered mechanism is triggered. Furthermore, the following inequality holds

$$|\varpi(t) - \upsilon(t)| \leq \varsigma(t)|\upsilon(t)| + \sigma.$$

Discuss the following two cases:

Case 1: $\upsilon(t) \geq 0$ leads to

$$\varpi(t) - \upsilon(t) = \delta(t)\Big(\varsigma(t)\upsilon(t) + \sigma\Big), \quad \delta(t) \in [-1, 1].$$

Case 2: $\upsilon(t) < 0$ brings about

$$\varpi(t) - \upsilon(t) = \delta(t)\Big(\varsigma(t)\upsilon(t) - \sigma\Big), \quad \delta(t) \in [-1, 1].$$

To sum up, we can see that

$$\varpi(t) - \upsilon(t) = \delta_1(t)\varsigma(t)\upsilon(t) + \delta_2(t)\sigma,$$

where

$$\begin{aligned} &\delta_1(t) = \delta_2(t) = \delta(t), \qquad && \upsilon(t) \geq 0 \\ &\delta_1(t) = \delta(t), \quad \delta_2(t) = -\delta(t), \quad && \upsilon(t) < 0. \end{aligned}$$

Then we can get

$$\upsilon(t) = \frac{\varpi(t)}{1 + \delta_1(t)\zeta(t)} - \frac{\delta_2(t)\sigma}{1 + \delta_1(t)\zeta(t)}, \tag{4.3}$$

where $|\delta_1(t)| \leq 1$ and $|\delta_2(t)| \leq 1$.

Remark 4.1 Ensuring system stability and avoiding the Zeno phenomenon are also the goals of event-triggered mechanism design. A good event-triggered mechanism can effectively reduce the communication frequency at the cost of only a small part of the performance loss. Compared with the fixed threshold event-triggered control method, the designed dynamic event-triggering mechanism has relatively fewer trigger times. Different from the relative threshold triggering mechanism used in Ma et al. (2020), the key feature of the dynamic event-triggering mechanism designed in this chapter is that the threshold parameter $\zeta(t)$ is not required to be a fixed constant value. As an alternative, it can be adjusted dynamically according to dynamic law $\dot{\zeta}(t) = -a\zeta^{\iota}(t)$. In particular, we define $\iota \geq 2$. This reduces the conservativeness, and the value of ι can be adjusted according to the control requirement.

4.2.3 *FLSs*

Since the nonlinear functions are unknown, the general control scheme with these functions cannot be implemented in practical applications. As recognized and effective means, FLSs and NNs are often adopted to approximate unknown nonlinear functions (Na et al. 2020; Zhang et al. 2021; Li et al. 2021, 2022; Liu et al. 2019; Qiu et al. 2020). FLSs are used to estimate the unknown nonlinearity, which are expressed as

Rule j**: IF** x_1 is $\bar{\mathfrak{D}}_{j1}$, x_2 is $\bar{\mathfrak{D}}_{j2}$, ..., and x_n is $\bar{\mathfrak{D}}_{jn}$, **THEN** y is $\bar{\mathfrak{E}}_j$, $j = 1, 2, \ldots, N$, where $\bar{\mathfrak{D}}_{jl}$ $(l = 1, 2, \ldots, n)$ and $\bar{\mathfrak{E}}_j$ denote fuzzy sets. Moreover, $x = [x_1, x_2, \ldots, x_n]^{\mathrm{T}}$ and y represent the input and output of FLSs, respectively.

The basic configuration of a FLS is shown in Fig. 4.1. Based on this, the FLS is described by

$$y(x) = \frac{\sum_{j=1}^{N} \bar{y}_j \prod_{l=1}^{n} \mu_{\bar{\mathfrak{D}}_{jl}}(x_l)}{\sum_{j=1}^{N} \prod_{l=1}^{n} \mu_{\bar{\mathfrak{D}}_{jl}}(x_l)}. \tag{4.4}$$

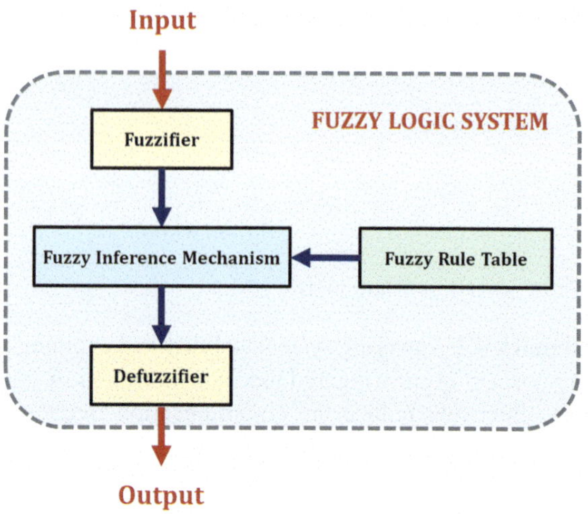

Fig. 4.1 Configuration of the FLS

Let $\bar{y}_j = \max_{y \in \mathbb{R}} \mu_{\bar{\mathfrak{E}}_j}(y)$, where $\mu_{\bar{\mathfrak{D}}_{jl}}(x_l)$ and $\mu_{\bar{\mathfrak{E}}_j}(y)$ are membership functions. The fuzzy basis functions are defined as

$$\varphi_j(x) = \frac{\prod_{l=1}^{n} \mu_{\bar{\mathfrak{D}}_{jl}}(x_l)}{\sum_{j=1}^{N} \prod_{l=1}^{n} \mu_{\bar{\mathfrak{D}}_{jl}}(x_l)}.$$

Then, the FLS (4.4) is rewritten as $y(x) = \theta^{\mathrm{T}} \varphi(x)$, where $\theta^{\mathrm{T}} = [\bar{y}_1, \bar{y}_2, \ldots, \bar{y}_N] = [\theta_1, \theta_2, \ldots, \theta_N]$, $\varphi(x) = [\varphi_1(x), \varphi_2(x), \ldots, \varphi_N(x)]^{\mathrm{T}}$. The fuzzy control lemma is described as follows.

Lemma 4.1 (Pan et al. 2021) *For the continuous function $\bar{f}(x)$ over a compact set Ω and any scalar $\bar{\epsilon} > 0$, there exists a FLS such that*

$$\sup_{x \in \Omega} \left| \bar{f}(x) - \theta^T \varphi(x) \right| \le \bar{\epsilon}.$$

4.3 Intelligent Controller Design and Stability Analysis

In this section, the design processes of the immediate control signals, actual controller, and adaptive laws are given, and the system performance is discussed as well. The block diagram of the adaptive dynamic event-triggered control scheme is shown in Fig. 4.2.

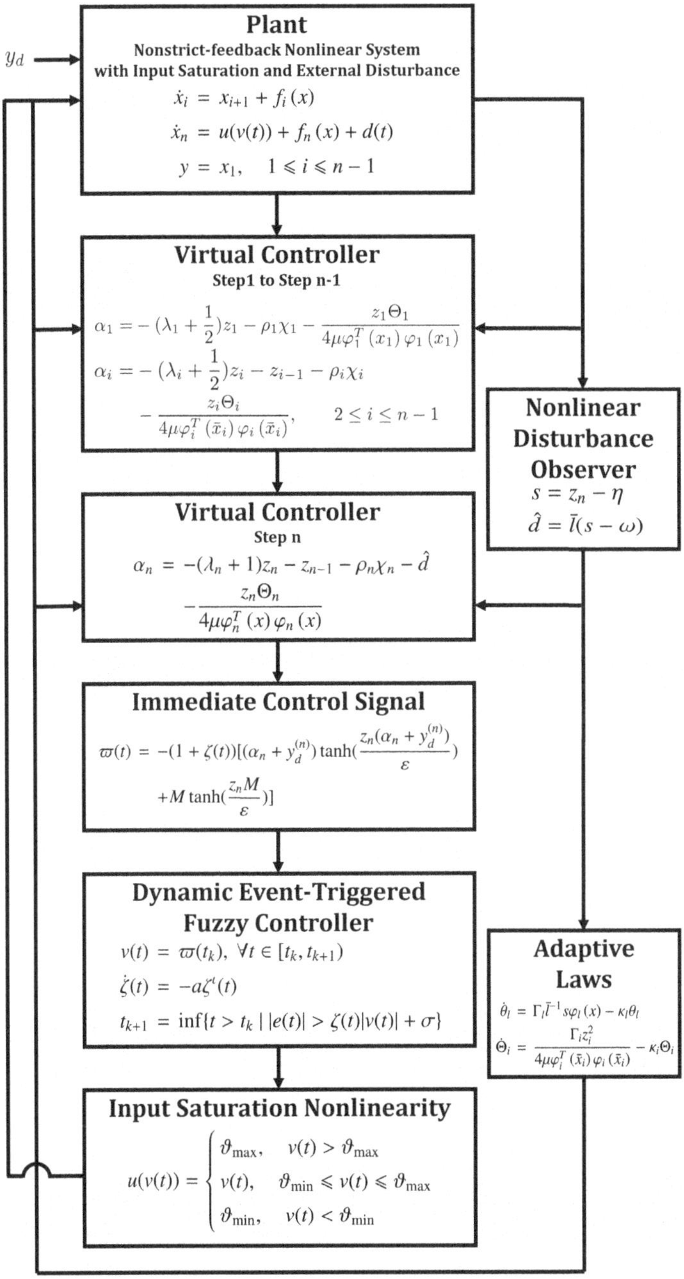

Fig. 4.2 Block diagram of the adaptive dynamic event-triggered control scheme

Firstly, for compensating the negative impact of input saturation on the considered system (4.1), design the following auxiliary system:

$$\begin{aligned}\dot{\chi}_i &= \chi_{i+1} - \rho_i \chi_i, \quad 1 \leq i \leq n-1 \\ \dot{\chi}_n &= \Delta u - \rho_n \chi_n,\end{aligned} \tag{4.5}$$

where the design parameters $\rho_1 = 1/2$ and $\rho_i \geq 1$ $(i = 2, 3, \ldots, n)$. Moreover, define $\Delta u = u(v(t)) - v(t)$.

For applying the backstepping technique, design the following coordinate transformations as

$$\begin{aligned}z_1 &= y - y_d - \chi_1 \\ z_i &= x_i - \alpha_{i-1} - y_d^{(i-1)} - \chi_i, \ i = 2, 3, \ldots, n,\end{aligned} \tag{4.6}$$

where y_d shows the desired tracking signal, and assume it and its ith derivatives are known and bounded. α_{i-1} denotes the virtual control signal.

Remark 4.2 Equation (4.6) is a more general form of coordinate transformation for $\chi_i = 0$ if $\chi_i(0) = 0$. When the saturation nonlinearity is not considered, the designed Eq. (4.5) becomes the traditional form as Tong et al. (2020), Ma et al. (2020). When the input saturation occurs, the auxiliary signal $\chi_i \neq 0$ if $\chi_i(0) \neq 0$. The changes in coordinates can also compensate the saturation nonlinearity regardless of the amplitude of reference signals.

Next, the recursive backstepping technique will be used to develop the adaptive fuzzy control laws. The virtual controllers α_i $(i = 1, 2, \ldots, n)$ are defined as

$$\begin{aligned}\alpha_1 &= -\left(\lambda_1 + \frac{1}{2}\right) z_1 - \rho_1 \chi_1 - \frac{z_1 \Theta_1}{4\mu \varphi_1^{\mathrm{T}}(x_1) \varphi_1(x_1)} \\ \alpha_i &= -\left(\lambda_i + \frac{1}{2}\right) z_i - z_{i-1} - \rho_i \chi_i \\ &\quad - \frac{z_i \Theta_i}{4\mu \varphi_i^{\mathrm{T}}(\bar{x}_i) \varphi_i(\bar{x}_i)}, \qquad 2 \leq i \leq n-1 \\ \alpha_n &= -(\lambda_n + 1) z_n - z_{n-1} - \rho_n \chi_n - \hat{d} - \frac{z_n \Theta_n}{4\mu \varphi_n^{\mathrm{T}}(x) \varphi_n(x)},\end{aligned} \tag{4.7}$$

where λ_i and μ are positive designed parameters. $\bar{x}_i = [x_1, x_2, \ldots, x_i]^{\mathrm{T}}$ $(i = 1, 2, \ldots, n)$. Specially, $\bar{x}_n = x$. $\hat{d}$ is the estimation of disturbance d. For $i = 1, 2, \ldots, n$, Θ_i and θ_l represent the estimations of Θ_i^* and θ_l^*, $\Theta_i^* = \left\|\theta_i^*\right\|^2$, and $\tilde{\Theta}_i = \Theta_i^* - \Theta_i$, $\tilde{\theta}_l = \theta_l^* - \theta_l$.

The adaptive laws $\dot{\theta}_l$ and $\dot{\Theta}_i$ $(i = 1, 2, \ldots, n)$ are designed as

$$\dot{\theta}_l = \Gamma_l \bar{l}^{-1} s \varphi_l(x) - \kappa_l \theta_l$$

$$\dot{\Theta}_i = \frac{\Gamma_i z_i^2}{4\mu\varphi_i^{\mathrm{T}}(\bar{x}_i)\varphi_i(\bar{x}_i)} - \kappa_i\Theta_i, \tag{4.8}$$

where Γ_l, Γ_i, κ_l and κ_i are positive constants to be designed.

In the following, the adaptive fuzzy controllers and corresponding event-triggering mechanism are co-designed for system (4.1). The main results are given by the following theorem.

Theorem 4.1 *For the system described by Eq.* (4.1) *with input saturation and external disturbance under Assumptions 4.1 and 4.2, the virtual controllers (4.7), the adaptive laws (4.8), the dynamic event-triggered mechanism (4.2), and the immediate control signal*

$$\varpi(t) = -\left(1+\zeta(t)\right)\left[(\alpha_n + y_d^{(n)})\tanh\left(\frac{z_n(\alpha_n + y_d^{(n)})}{\varepsilon}\right) + M\tanh\left(\frac{z_n M}{\varepsilon}\right)\right], \tag{4.9}$$

where $M > \frac{\sigma}{1-\zeta(t)}$ and ε is a positive design parameter, the boundedness of all signals in the resulting closed-loop system can be guaranteed, and the tracking error remains in a small neighborhood of the origin.

Proof The proof is consisted of n steps.

Step 1: The first change of coordinate is $z_1 = y - y_d - \chi_1$, and its time derivative along the first subsystems of (4.1) and (4.5) is given by

$$\dot{z}_1 = x_2 + f_1(x) - \dot{y}_d - \chi_2 + \rho_1\chi_1. \tag{4.10}$$

The FLS is employed to approximate the unknown nonlinear smooth function $f_1(x)$. Then, we have

$$f_1(x) = \theta_1^{*\mathrm{T}}\varphi_1(x) + \epsilon_1(x), \quad |\epsilon_1(x)| \le \bar{\epsilon}_1.$$

For handling the difficulties arising from the nonstrict-feedback structure of the system (4.1), another parameter Θ_1^* is defined as $\Theta_1^* = \left\|\theta_1^*\right\|^2$. Moreover, invoking $0 < \varphi_1^{\mathrm{T}}(\cdot)\varphi_1(\cdot) \le 1$, the following inequality can be obtained:

$$\begin{aligned} z_1\theta_1^{*\mathrm{T}}\varphi_1(x) &\le \frac{z_1^2\theta_1^{*\mathrm{T}}\theta_1^*\varphi_1^{\mathrm{T}}(x)\varphi_1(x)}{4\mu} + \mu \\ &\le \frac{z_1^2\Theta_1^*}{4\mu\varphi_1^{\mathrm{T}}(x_1)\varphi_1(x_1)} + \mu. \end{aligned} \tag{4.11}$$

Consider the Lyapunov function candidate

$$V_1 = \frac{1}{2}z_1^2 + \frac{1}{2\Gamma_1}\tilde{\Theta}_1^2.$$

In view of Eqs. (4.10) and (4.11), differentiating V_1 yields

$$\begin{aligned}\dot{V}_1 \leq\ & z_1\Bigg[z_2 + \alpha_1 + \frac{z_1\Theta_1^*}{4\mu\varphi_1^{\mathrm{T}}(x_1)\varphi_1(x_1)} + \rho_1\chi_1 \\ & + \epsilon_1(x)\Bigg] - \frac{1}{\Gamma_1}\tilde{\Theta}_1\dot{\Theta}_1 + \mu.\end{aligned}$$

Based on Young's inequality, we get

$$z_1\epsilon_1(x) \leq \frac{1}{2}z_1^2 + \frac{1}{2}\bar{\epsilon}_1^2.$$

Therefore, one has

$$\begin{aligned}\dot{V}_1 \leq\ & z_1\left[z_2 + \alpha_1 + \frac{z_1\Theta_1}{4\mu\varphi_1^{\mathrm{T}}(x_1)\varphi_1(x_1)} + \rho_1\chi_1 + \frac{1}{2}z_1\right] \\ & + \frac{1}{\Gamma_1}\tilde{\Theta}_1\left[\frac{\Gamma_1 z_1^2}{4\mu\varphi_1^{\mathrm{T}}(x_1)\varphi_1(x_1)} - \dot{\Theta}_1\right] + \mu + \frac{1}{2}\bar{\epsilon}_1^2.\end{aligned}$$

Together with the first virtual control signal α_1 and adaptive law $\dot{\Theta}_1$, the derivative of V_1 is given as

$$\dot{V}_1 \leq -\lambda_1 z_1^2 + z_1 z_2 + \frac{\kappa_1}{\Gamma_1}\tilde{\Theta}_1\Theta_1 + \mu + \frac{1}{2}\bar{\epsilon}_1^2. \tag{4.12}$$

The following formula (4.13) holds via Young's inequality

$$\frac{\kappa_1}{\Gamma_1}\tilde{\Theta}_1\Theta_1 \leq -\frac{\kappa_1}{2\Gamma_1}\tilde{\Theta}_1^2 + \frac{\kappa_1}{2\Gamma_1}\Theta_1^{*2}. \tag{4.13}$$

Finally, in Step 1, the inequality (4.12) can be modified as

$$\dot{V}_1 \leq -\lambda_1 z_1^2 + z_1 z_2 - \frac{\kappa_1}{2\Gamma_1}\tilde{\Theta}_1^2 + B_1,$$

where $B_1 = \frac{\kappa_1}{2\Gamma_1}\Theta_1^{*2} + \mu + \frac{1}{2}\bar{\epsilon}_1^2$.

Step i ($i = 2, 3, \ldots, n-1$): By considering Eqs. (4.1), (4.5) and (4.6), differentiating z_i with respect to time yields

$$\dot{z}_i = x_{i+1} + f_i(x) - \dot{\alpha}_{i-1} - y_d^{(i)} - \chi_{i+1} + \rho_i\chi_i.$$

The following inequality can be obtained by applying Lemma 4.1:

$$f_i(x) - \dot{\alpha}_{i-1} = \theta_i^{*\mathrm{T}} \varphi_i(x) + \epsilon_i(x), \ |\epsilon_i(x)| \le \bar{\epsilon}_i.$$

Then, we can obtain

$$z_i \theta_i^{*\mathrm{T}} \varphi_i(x) \le \frac{z_i^2 \Theta_i^*}{4\mu \varphi_i^{\mathrm{T}}(\bar{x}_i)\varphi_i(\bar{x}_i)} + \mu.$$

Define the Lyapunov function as follows:

$$V_i = V_{i-1} + \frac{1}{2} z_i^2 + \frac{1}{2\Gamma_i} \tilde{\Theta}_i^2. \tag{4.14}$$

Based on the same method as Step 1, taking the time derivatives of Eq. (4.14) results in

$$\begin{aligned}\dot{V}_i \le & -\sum_{j=1}^{i-1} \lambda_j z_j^2 - \sum_{j=1}^{i-1} \frac{\kappa_j}{2\Gamma_j} \tilde{\Theta}_j^2 + B_{i-1} + z_i \Big[z_{i+1} + \alpha_i \\ & + z_{i-1} + \frac{z_i \Theta_i}{4\mu \varphi_i^{\mathrm{T}}(\bar{x}_i)\varphi_i(\bar{x}_i)} + \rho_i \chi_i + \epsilon_i(x) \Big] \\ & + \frac{1}{\Gamma_i} \tilde{\Theta}_i \left[\frac{\Gamma_i z_i^2}{4\mu \varphi_i^{\mathrm{T}}(\bar{x}_i)\varphi_i(\bar{x}_i)} - \dot{\Theta}_i \right] + \mu, \end{aligned} \tag{4.15}$$

where $B_{i-1} = (i-1)\mu + \sum_{j=1}^{i-1} \frac{\kappa_j}{2\Gamma_j} \Theta_j^{*2} + \sum_{j=1}^{i-1} \frac{1}{2} \bar{\epsilon}_j^2$. According to Young's inequality, one has

$$z_i \epsilon_i(x) \le \frac{1}{2} z_i^2 + \frac{1}{2} \bar{\epsilon}_i^2.$$

Substituting Eqs. (4.7) and (4.8) into the inequality (4.15) yields

$$\begin{aligned}\dot{V}_i \le & -\sum_{j=1}^{i} \lambda_j z_j^2 - \sum_{j=1}^{i-1} \frac{\kappa_j}{2\Gamma_j} \tilde{\Theta}_j^2 + B_{i-1} + \mu + \frac{1}{2} \bar{\epsilon}_i^2 \\ & + z_i z_{i+1} + \frac{\kappa_i}{\Gamma_i} \tilde{\Theta}_i \Theta_i. \end{aligned} \tag{4.16}$$

By using the same approach as Step 1, the following formula is derived:

$$\frac{\kappa_i}{\Gamma_i} \tilde{\Theta}_i \Theta_i \le -\frac{\kappa_i}{2\Gamma_i} \tilde{\Theta}_i^2 + \frac{\kappa_i}{2\Gamma_i} \Theta_i^{*2}.$$

Then, Eq. (4.16) can be rewritten as

$$\dot{V}_i \leq -\sum_{j=1}^{i} \lambda_j z_j^2 - \sum_{j=1}^{i} \frac{\kappa_j}{2\Gamma_j} \tilde{\Theta}_j^2 + B_i + z_i z_{i+1},$$

where $B_i = B_{i-1} + \frac{\kappa_i}{2\Gamma_i}\Theta_i^{*2} + \mu + \frac{1}{2}\bar{\epsilon}_i^2$.

Step *n*: Similar to the above steps, $\dot{z}_n$ is given as

$$\dot{z}_n = \upsilon(t) + f_n(x) + d(t) - \dot{\alpha}_{n-1} - y_d^{(n)} + \rho_n \chi_n.$$

Consider a Lyapunov function candidate as

$$V_n = V_{n-1} + \frac{1}{2} z_n^2 + \frac{1}{2\Gamma_n} \tilde{\Theta}_n^2 + \frac{1}{2} \tilde{d}^2 + \frac{1}{2} s^2 + \frac{1}{2\Gamma_l} \tilde{\theta}_l^2, \tag{4.17}$$

where $\tilde{d} = d - \hat{d}$ and $\hat{d}$ is the estimation of d. Moreover, s is an auxiliary variable, which will be defined later. Taking the time derivative of Eq. (4.17) gives

$$\begin{aligned}
\dot{V}_n \leq & -\sum_{i=1}^{n-1} \lambda_i z_i^2 - \sum_{i=1}^{n-1} \frac{\kappa_i}{2\Gamma_i} \tilde{\Theta}_i^2 + B_{n-1} + z_{n-1} z_n \\
& + z_n \Big[\upsilon(t) + f_n(x) + d(t) - \dot{\alpha}_{n-1} - y_d^{(n)} + \rho_n \chi_n \Big] \\
& - \frac{1}{\Gamma_n} \tilde{\Theta}_n \dot{\Theta}_n + \tilde{d}\dot{\tilde{d}} + s\dot{s} - \frac{1}{\Gamma_l} \tilde{\theta}_l \dot{\theta}_l.
\end{aligned}$$

Followed by Eq. (4.3), we have

$$\begin{aligned}
\dot{V}_n \leq & -\sum_{i=1}^{n-1} \lambda_i z_i^2 - \sum_{i=1}^{n-1} \frac{\kappa_i}{2\Gamma_i} \tilde{\Theta}_i^2 + B_{n-1} + z_{n-1} z_n \\
& - \frac{1}{\Gamma_n} \tilde{\Theta}_n \dot{\Theta}_n + \tilde{d}\dot{\tilde{d}} + s\dot{s} - \frac{1}{\Gamma_l} \tilde{\theta}_l \dot{\theta}_l + z_n \Big[\frac{\varpi(t)}{1 + \delta_1(t)\varsigma(t)} - \frac{\delta_2(t)\sigma}{1 + \delta_1(t)\varsigma(t)} \\
& + f_n(x) + d(t) - \dot{\alpha}_{n-1} - y_d^{(n)} + \rho_n \chi_n \Big].
\end{aligned}$$

By the designed immediate control signal, it yields

$$\begin{aligned}
\dot{V}_n \leq & -\sum_{i=1}^{n-1} \lambda_i z_i^2 - \sum_{i=1}^{n-1} \frac{\kappa_i}{2\Gamma_i} \tilde{\Theta}_i^2 + B_{n-1} + z_{n-1} z_n \\
& - \frac{1}{\Gamma_n} \tilde{\Theta}_n \dot{\Theta}_n + \tilde{d}\dot{\tilde{d}} + s\dot{s} - \frac{1}{\Gamma_l} \tilde{\theta}_l \dot{\theta}_l
\end{aligned}$$

$$
\begin{aligned}
&+z_n\Big[-(\alpha_n+y_d^{(n)})\tanh\left(\frac{z_n(\alpha_n+y_d^{(n)})}{\varepsilon}\right)\\
&-M\tanh\left(\frac{z_nM}{\varepsilon}\right)-\frac{\sigma}{1-\zeta(t)}\\
&+f_n(x)+d(t)-\dot{\alpha}_{n-1}-y_d^{(n)}+\rho_n\chi_n\Big].
\end{aligned}
\tag{4.18}
$$

By using $0\le|\varkappa|-\varkappa\tanh(\frac{\varkappa}{\varepsilon})\le 0.2785\varepsilon$, then we can get

$$
\begin{aligned}
&-z_n(\alpha_n+y_d^{(n)})\tanh\left(\frac{z_n(\alpha_n+y_d^{(n)})}{\varepsilon}\right)\\
\le\ &|z_n(\alpha_n+y_d^{(n)})|+0.2785\varepsilon
\end{aligned}
\tag{4.19}
$$

and

$$
-z_nM\tanh\left(\frac{z_nM}{\varepsilon}\right)\le-|z_nM|+0.2785\varepsilon. \tag{4.20}
$$

Substituting the inequalities (4.19) and (4.20) into (4.18), we obtain

$$
\begin{aligned}
\dot{V}_n\le\ &-\sum_{i=1}^{n-1}\lambda_iz_i^2-\sum_{i=1}^{n-1}\frac{\kappa_i}{2\Gamma_i}\tilde{\Theta}_i^2+B_{n-1}+\frac{1}{2}\bar{\epsilon}_n^2+\mu\\
&+z_n\Big[z_{n-1}+\alpha_n+d(t)+\frac{1}{2}z_n^2+\rho_n\chi_n\\
&+\frac{z_n\Theta_n}{4\mu\varphi_n^{\mathrm{T}}(x)\varphi_n(x)}\Big]+\tilde{d}\dot{\tilde{d}}+s\dot{s}-\frac{1}{\Gamma_l}\tilde{\theta}_l\dot{\theta}_l\\
&+\frac{1}{\Gamma_n}\tilde{\Theta}_n\left[\frac{\Gamma_nz_n^2}{4\mu\varphi_n^{\mathrm{T}}(x)\varphi_n(x)}-\dot{\Theta}_n\right].
\end{aligned}
$$

By designing virtual controller α_n and adaptive law $\dot{\Theta}_n$, it yields

$$
\dot{V}_n\le-\sum_{i=1}^{n}\lambda_iz_i^2-\sum_{i=1}^{n}\frac{\kappa_i}{2\Gamma_i}\tilde{\Theta}_i^2+B_n+\frac{1}{2}\tilde{d}+\tilde{d}\dot{\tilde{d}}+s\dot{s}-\frac{1}{\Gamma_l}\tilde{\theta}_l\dot{\theta}_l,
$$

where $B_n=B_{n-1}+\frac{1}{2}\bar{\epsilon}_n^2+\mu+\frac{\kappa_l}{2\Gamma_l}\Theta_n^{*2}$. To facilitate the design of the nonlinear disturbance observer, $\dot{z}_n$ can be expressed as

$$
\dot{z}_n=\bar{l}^{-1}g(x,v,y_d)+d(t)-\dot{\alpha}_{n-1}+\rho_n\chi_n,
$$

where $g(x,v,y_d)=\bar{l}[f_n(x)+v(t)-y_d^{(n)}]$ and $\bar{l}$ is a positive constant of the developed nonlinear disturbance observer. There exists a FLS $\theta_l^{*\mathrm{T}}\varphi_l(x)$ such that

$$\dot{z}_n = \bar{l}^{-1}\theta_l^{*\mathrm{T}}\varphi_l(x) + \bar{l}^{-1}\epsilon_l + d(t) - \dot{\alpha}_{n-1} + \rho_n\chi_n.$$

Define the following auxiliary variable:

$$s = z_n - \eta$$

with the intermediate variable η as

$$\dot{\eta} = \bar{c}s + \bar{l}^{-1}\theta_l^{\mathrm{T}}\varphi_l(x) - \dot{\alpha}_{n-1} + \rho_n\chi_n,$$

where $\bar{c} > 0$. For $\tilde{\theta}_l = \theta_l^* - \theta_l$, the time derivative of s is

$$\dot{s} = \bar{l}^{-1}\tilde{\theta}_l^{\mathrm{T}}\varphi_l(x) + \bar{l}^{-1}\epsilon_l + d(t) - \bar{c}s.$$

By using Young's inequality, we have

$$s\dot{s} \leq -(\bar{c} - 1)s^2 + \bar{l}^{-1}s\tilde{\theta}_l^{\mathrm{T}}\varphi_l(x) + \frac{1}{2}\bar{l}^{-2}\bar{\epsilon}_l^2 + \frac{1}{2}\bar{d}_0^2.$$

Based on the auxiliary variable s, the nonlinear disturbance observer is designed as

$$\hat{d} = \bar{l}(s - \omega)$$

with the intermedial variable ω as

$$\dot{\omega} = -\bar{c}s + \hat{d}.$$

Differentiating $\hat{d}$ with respect to time leads to

$$\dot{\hat{d}} = \bar{l}\tilde{d} + \tilde{\theta}_l^{\mathrm{T}}\varphi_l(x) + \epsilon_l$$

and

$$\tilde{d}\dot{\tilde{d}} = \tilde{d}\dot{d} - \bar{l}\tilde{d}^2 - \tilde{d}\tilde{\theta}_l^{\mathrm{T}}\varphi_l(x) - \tilde{d}\epsilon_l.$$

Consider the following fact as

$$-\tilde{d}\tilde{\theta}_l^{\mathrm{T}}\varphi_l(x) \leq \frac{1}{2}\varrho M^2\tilde{d}^2 + \frac{1}{2\varrho}||\tilde{\theta}_l||^2,$$

where $||\varphi_l(x)|| \leq M$. M and ϱ are the positive design parameters. Based on Young's inequality, one has

$$\tilde{d}\dot{\tilde{d}} \leq \tilde{d}^2 + \frac{1}{2}\bar{d}_1^2 - \bar{l}\tilde{d}^2 + \frac{1}{2}\varrho M^2\tilde{d}^2 + \frac{1}{2\varrho}||\tilde{\theta}_l||^2 + \frac{1}{2}\bar{\epsilon}_l^2.$$

Then, $\dot{V}_n$ can be rewritten as

$$\begin{aligned}\dot{V}_n \le & -\sum_{i=1}^{n}\lambda_i z_i^2 - \sum_{i=1}^{n}\frac{\kappa_i}{2\Gamma_i}\tilde{\Theta}_i^2 + B_n + \frac{3}{2}\tilde{d}^2 - \frac{1}{\Gamma_l}\tilde{\theta}_l\dot{\theta}_l \\ & + \frac{1}{2}\bar{d}_1^2 - \bar{l}\tilde{d}^2 + \frac{1}{2}\varrho M^2\tilde{d}^2 + \frac{1}{2\varrho}\|\tilde{\theta}_l\|^2 + \frac{1}{2}\bar{\epsilon}_l^2 \\ & -(\bar{c}-1)s^2 + \bar{l}^{-1}s\tilde{\theta}_l^{\mathrm{T}}\varphi_l(x) + \frac{1}{2}\bar{l}^{-2}\bar{\epsilon}_l^2 + \frac{1}{2}\bar{d}_0^2.\end{aligned}$$

By designing the adaptive law $\dot{\theta}_l = \Gamma_l\bar{l}^{-1}s\varphi_l(x) - \kappa_l\theta_l$, we have

$$\begin{aligned}\dot{V}_n \le & -\sum_{i=1}^{n}\lambda_i z_i^2 - \sum_{i=1}^{n}\frac{\kappa_i}{2\Gamma_i}\tilde{\Theta}_i^2 + \bar{B}_l - \left(\bar{l} - \frac{3}{2} - \frac{1}{2}\varrho M^2\right)\tilde{d}^2 - (\bar{c}-1)s^2 \\ & -\left(\frac{\kappa_l}{2\Gamma_l} - \frac{1}{2\varrho}\right)\tilde{\theta}_l{}^2,\end{aligned}$$

where $\bar{B}_l = B_n + \frac{1}{2}(\bar{d}_1^2 + \bar{d}_0^2 + \bar{\epsilon}_l^2 + \bar{l}^{-2}\bar{\epsilon}_l^2 + \frac{\kappa_l}{2\Gamma_l}\theta_l^{*2})$. Finally, the following expression holds

$$\dot{V}_n \le -\bar{A}_l V_n + \bar{B}_l, \tag{4.21}$$

where $\bar{A}_l = \min\{2\lambda_i, \kappa_i, 2\bar{l} - 3 - \varrho M^2, 2\bar{c} - 2, \kappa_l - \frac{\Gamma_l}{\varrho}, i = 1, 2, \ldots, n\}$. To ensure the stability of the closed-loop system, the corresponding design parameters should be chosen to make the following inequalities hold

$$\begin{aligned}\bar{l} - \frac{3}{2} - \frac{1}{2}\varrho M^2 &> 0 \\ \bar{c} - 1 &> 0 \\ \frac{\kappa_l}{2\Gamma_l} - \frac{1}{2\varrho} &> 0.\end{aligned}$$

From Eq. (4.21), we can get

$$0 \le V_n \le \frac{\bar{B}_l}{\bar{A}_l} + \left[V_n(0) - \frac{\bar{B}_l}{\bar{A}_l}\right]e^{-\bar{A}_l t}. \tag{4.22}$$

The inequality (4.22) indicates that V_n is convergent. For this inequality, it follows that

$$|z_1| \le \sqrt{\frac{2\bar{B}_l}{\bar{A}_l} + 2\left[V_n(0) - \frac{\bar{B}_l}{\bar{A}_l}\right]e^{-\bar{A}_l t}}.$$

Consider the following fact as

$$|y - y_d|^2 \leq 2z_1^2 + 2\chi_1^2.$$

To derive the bound of $|\chi_1|$, construct the Lyapunov function candidate as

$$V_\chi = \frac{1}{2}\sum_{i=1}^{n} \chi_i^2.$$

Based on Eq. (4.5), $\dot{V}_\chi$ is calculated as

$$\dot{V}_\chi \leq -\bar{P}V_\chi + \frac{1}{2}(\Delta u)^2,$$

where $\bar{P} = \min\{2P_i, i = 1, 2, \ldots, n\}$ and $P_1 = \rho_1 - 1/2$, $P_i = \rho_i - 1$ $(i = 2, 3, \ldots, n)$. Then, we can get the bound of $|\chi_1|$ as

$$|\chi_1| \leq \sqrt{\frac{\sup\limits_{0\leq\bar{s}\leq t} |\Delta u(\bar{s})|^2}{\bar{P}} + \left[2V_\chi(0) - \frac{\sup\limits_{0\leq\bar{s}\leq t} |\Delta u(\bar{s})|^2}{\bar{P}}\right] e^{-\bar{P}t}}.$$

From Assumption 4.1, it can be seen that Δu is bounded. Therefore, one obtains that the tracking error remains in a neighborhood of zero by adjusting the corresponding parameters. In addition, all closed-loop signals are bounded through the above analysis. The proof is completed. □

In the following description, we show that the Zeno behavior is excluded effectively. There exists a positive constant t^* satisfying $\{t_{k+1} - t_k\} \geq t^*$, $\forall k \in Z^+$. Since $e(t) = \varpi(t) - v(t)$, $\forall t \in [t_k, t_{k+1})$, one has $\frac{d}{dt}|e(t)| \leq |\dot{\varpi}(t)|$. According to the control law (4.9), it can be known that ϖ is differentiable. The inequality $|\dot{\varpi}(t)| < \Delta$ holds with Δ being a positive constant. The Zeno behavior is excluded successfully when $t^* \geq (\varsigma(t)v(t) + \sigma)/\Delta$.

Remark 4.3 In traditional control methods, data transmission is completely determined based on time. However, time-triggered methods usually result in redundant transmissions. Because all information is not necessarily required to maintain the control performance of a controlled system. For example, when the state of the controlled system tends to be stable, the measurement data at this time has a relatively small effect on the system, and the continued transmission of such information will cause a relative waste of network resources. Therefore, the event-triggered control method can balance resource utilization and control performance better than time-triggered control strategy.

4.4 Simulation Results

In this section, two simulation examples are given to show the effectiveness of the developed adaptive dynamic event-triggered control strategy in this chapter. One is a numerical example, and the other is a practical example of manipulator. Different predetermined tracking signals and disturbances are given for two examples, respectively.

Example 4.1 Consider the nonstrict-feedback nonlinear system in the form of model (4.1) with input saturation and external disturbance as

$$\begin{aligned} \dot{x}_1 &= x_2 + 0.5\sin(x_1 + x_2) \\ \dot{x}_2 &= u(v(t)) + \cos(x_2) + \sin(x_1) + d(t) \\ y &= x_1, \end{aligned}$$

where $d(t) = 0.2\sin(3t+1) + 0.1\cos(t)$. Define the desired tracking signal as $y_d(t) = \sin(2t) + 2\cos(t)$. The parameters of the designed adaptive dynamic event-triggered controller are chosen as $\rho_1 = 0.5,\ \rho_2 = 40,\ \lambda_1 = 25,\ \lambda_2 = 29,\ \mu = 26,\ \Gamma_l = 82,\ \Gamma_1 = 98,\ \Gamma_2 = 73,\ \kappa_l = 58,\ \kappa_1 = 10,\ \kappa_2 = 90,\ \bar{l} = 0.6,\ \bar{c} = 10\, a = 59,\ \sigma = 2,\ M = 4,\ \epsilon = 16,\ \vartheta_{\max} = 10,\ \vartheta_{\min} = -10$.

Within the simulation results, Fig. 4.3 shows the bounded trajectories of y and y_d and the tracking error. Figure 4.4 indicates that the trajectory of state x_2. In Fig. 4.5, the adaptive parameters are plotted. Figure 4.6 shows the trigger intervals between adjacent events. Figure 4.7 depicts that the system input saturation u and the saturation control input v are bounded as well. It is clearly seen that under the action of the proposed control law, all the closed-loop signals are semi-globally bounded.

Example 4.2 Based on Chen et al. (2015), Li et al. (2018, 2020), a single-link manipulator with the inclusion of motor dynamics, input saturation and external disturbance is considered as

$$\begin{aligned} \breve{I}\ddot{q}_l + \breve{N}\sin q_l + \breve{C}\dot{q}_l &= \tau \\ \breve{H}\tau + \breve{J}\dot{\tau} &= u(v(t)) + d(t) - \breve{K}\dot{q}_l, \end{aligned}$$

where $u(v(t))$ is the control input, and it is used to express electromechanical torque. Define the input saturation limit as $\vartheta_{\max} = -\vartheta_{\min} = 100\,\text{N m}$. τ represents the torque produced by the electrical subsystem. q_l, $\dot{q}_l$ and $\ddot{q}_l$ denote the link position, velocity and acceleration, respectively. $\breve{I} = 1\,\text{kg m}^2$ and $\breve{J} = 0.05\text{H}$ are the mechanical inertia and armature inductance, respectively. $\breve{N} = 10$ is a position constant related to the mass of the load and the coefficient of gravity. $\breve{H} = 10\,\Omega$ is the armature resistance. $\breve{C} = 1\,\text{N m s/rad}$ and $\breve{K} = 10\,\text{N m/A}$ show the coefficients of viscous friction at joint and the back electromotive force, respectively.

Let $x_1 = q_l,\ x_2 = \dot{q}_l$, and $x_3 = \ddot{q}_l$, the single-link manipulator system can be expressed in the form of Eq. (4.1) as

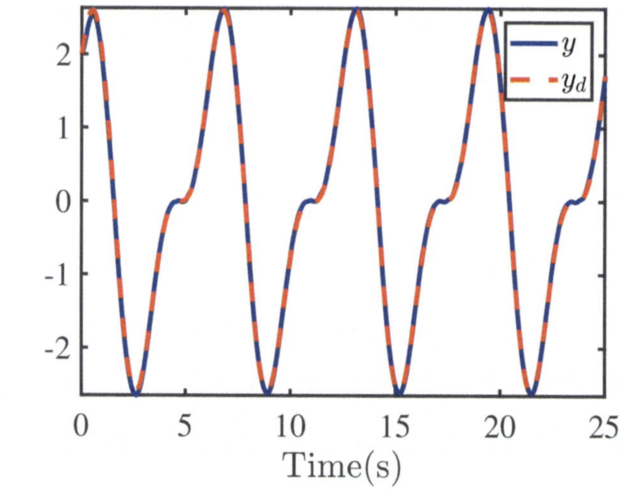

(a) Responses of y and y_d.

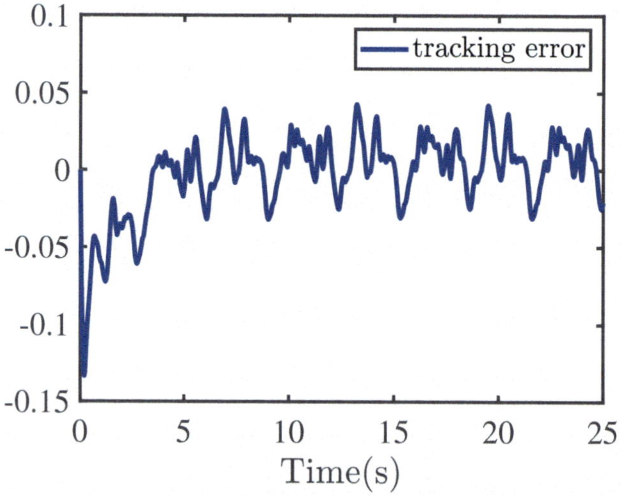

(b) Response of tracking error.

Fig. 4.3 System tracking effect in Example 4.1

$$
\begin{aligned}
\dot{x}_1 &= x_2, \\
\dot{x}_2 &= \left[x_3 - \breve{C}x_2 - \breve{N}\sin(x_1)\right]/\breve{I}, \\
\dot{x}_3 &= \left[u(v(t)) - \breve{K}x_2 - \breve{H}x_3 + d(t)\right]/\breve{J}, \\
y &= x_1,
\end{aligned}
$$

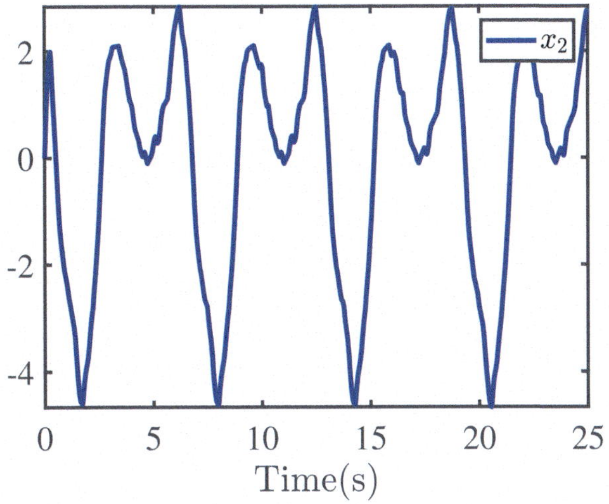

Fig. 4.4 Response of x_2 in Example 4.1

where $d(t) = 0.4\cos(2t)$. Define the desired tracking signal as $y_d(t) = \sin(t)$. The parameters of the designed adaptive dynamic event-triggered controller are chosen as $\rho_1 = 0.5$, $\rho_2 = 80$, $\rho_3 = 20$, $\lambda_1 = 35$, $\lambda_2 = 39$, $\lambda_3 = 6$, $\mu = 6$, $\Gamma_l = 82$, $\Gamma_1 = 98$, $\Gamma_2 = 73$, $\Gamma_3 = 34$, $\kappa_l = 58$, $\kappa_1 = 10$, $\kappa_2 = 90$, $\kappa_3 = 88$, $\bar{l} = 0.4$, $\bar{c} = 0.4$ $a = 59$, $\sigma = 2$, $M = 4$, $\epsilon = 16$.

The simulation results are shown in Figs. 4.8, 4.9, 4.10, 4.11, and 4.12 for the single-link manipulator. The responses of y and y_d are shown in Fig. 4.8a and the tracking error is shown in Fig. 4.8b. The trajectories of states x_2 and x_3 are shown in Fig. 4.9. Figure 4.10a plots the responses of adaptive parameters Θ_1, Θ_2, and Θ_3. The response of $\|\theta_l\|$ is revealed in Fig. 4.10b. The time intervals of every two adjacent trigger instants for Example 4.2 are plotted in Fig. 4.11. In addition, Fig. 4.12 depicts the trajectories of signals v and u. Based on the above simulation results, we can get that the developed fuzzy control method is effective.

4.5 Conclusion

A dynamic event-triggered control strategy has been proposed for a class of nonstrict-feedback nonlinear systems, it can avoid the Zeno behavior and guarantee closed-loop system stability simultaneously. In order to make better use of bandwidth resources, the dynamic event-triggered mechanism has been designed to reduce data transmission from controller to actuator. In the framework of adaptive backstepping method, a disturbance observer-based adaptive fuzzy control scheme has been proposed using the fuzzy basis function property of $0 < \varphi_1^{\mathrm{T}}(\cdot)\varphi_1(\cdot) \leq 1$, which can guarantee that all the signals are bounded in the system with input saturation. Finally, simulation

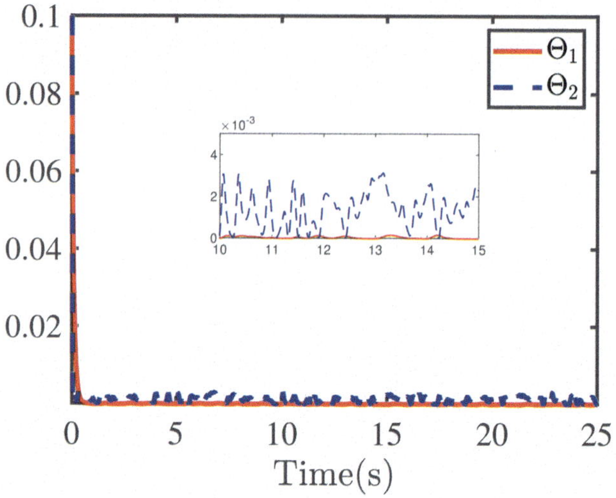

(a) Θ_1 and Θ_2.

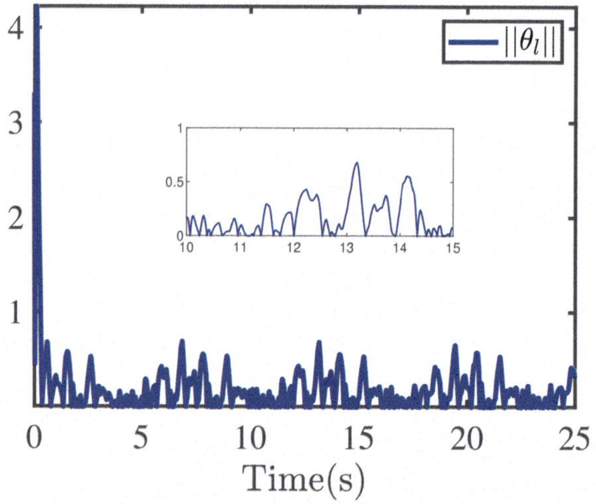

(b) $||\theta_l||$.

Fig. 4.5 Responses of adaptive parameters in Example 4.1

results have been given to verify the effectiveness of the control strategy proposed in this chapter. In the future research work, we will strive to apply the method to more practical systems (Wang et al. 2021; Han et al. 2021; Xia and Cheng 2021).

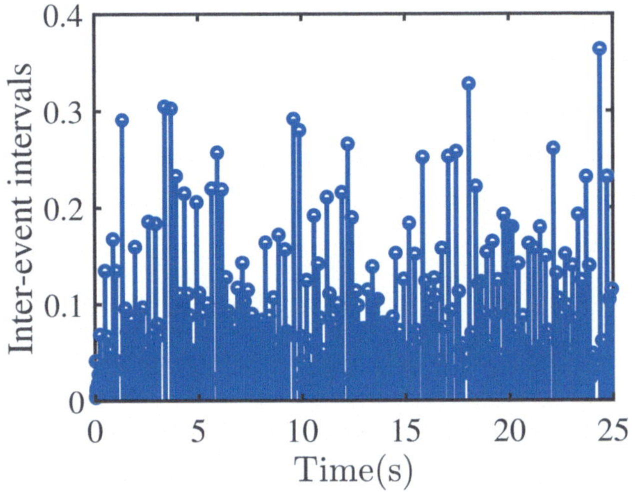

Fig. 4.6 (Color online) Trigger intervals between adjacent events in Example 4.1

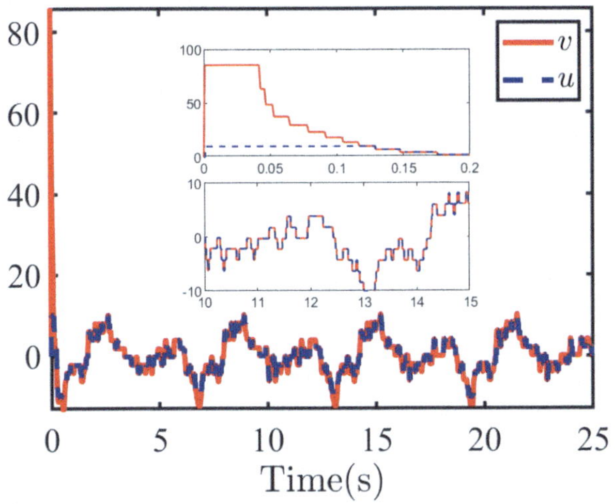

Fig. 4.7 Responses of v and u in Example 4.1

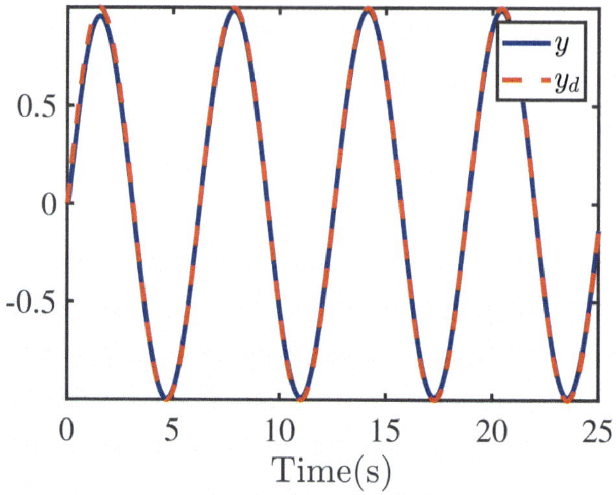

(a) Responses of y and y_d.

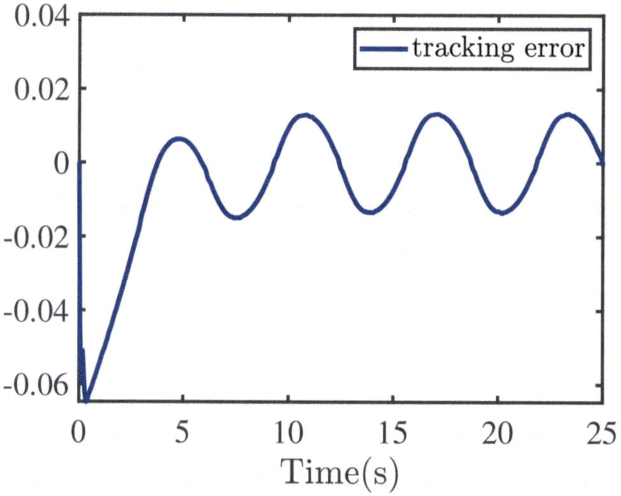

(b) Response of tracking error.

Fig. 4.8 System tracking effect in Example 4.2

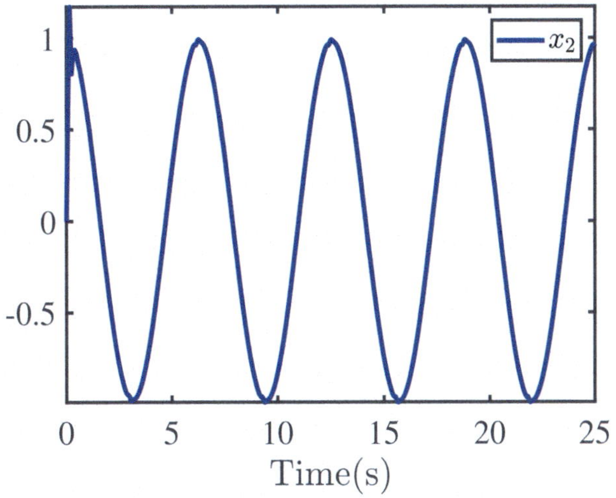

(a) Response of x_2.

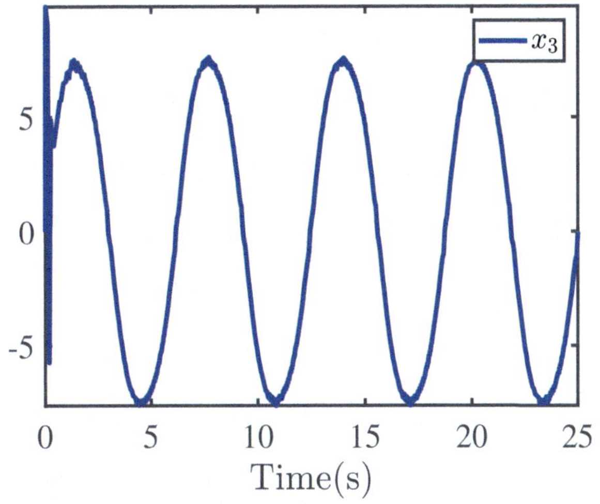

(b) Response of x_3.

Fig. 4.9 Other system states in Example 4.2

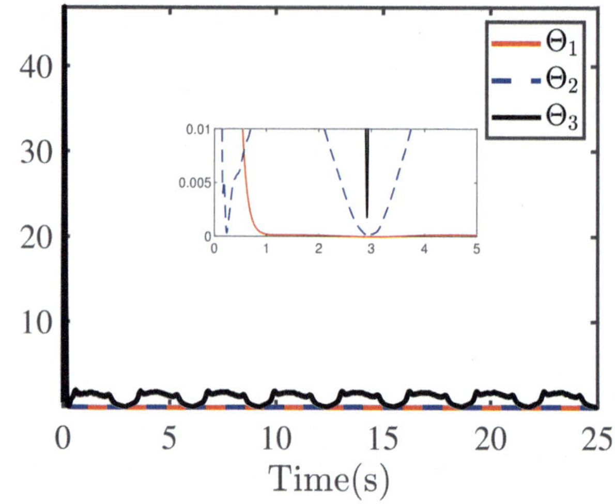

(a) Θ_1, Θ_2 and Θ_3.

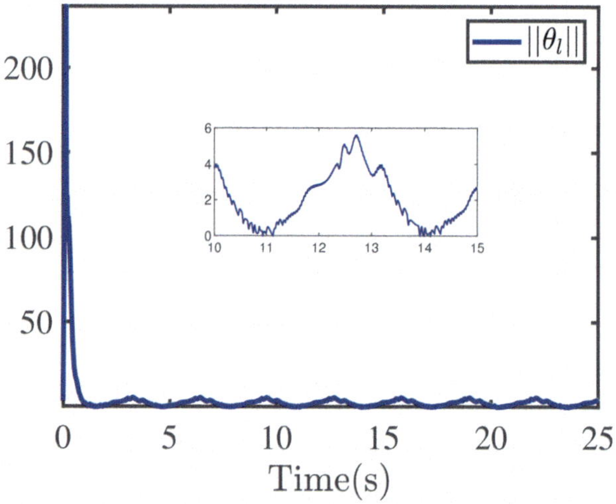

(b) $||\theta_l||$.

Fig. 4.10 Responses of adaptive parameters in Example 4.2

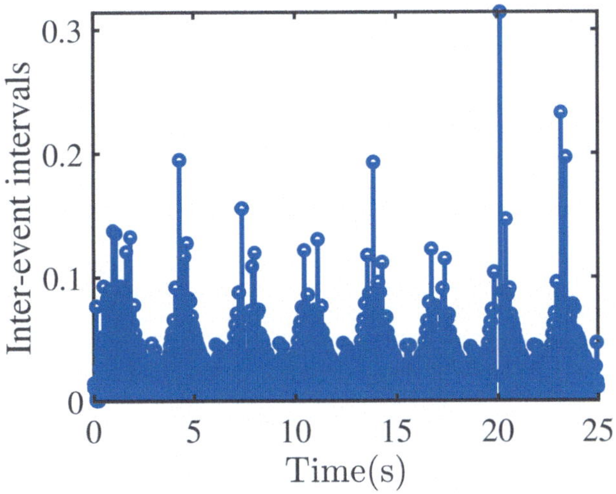

Fig. 4.11 Trigger intervals between adjacent events in Example 4.2

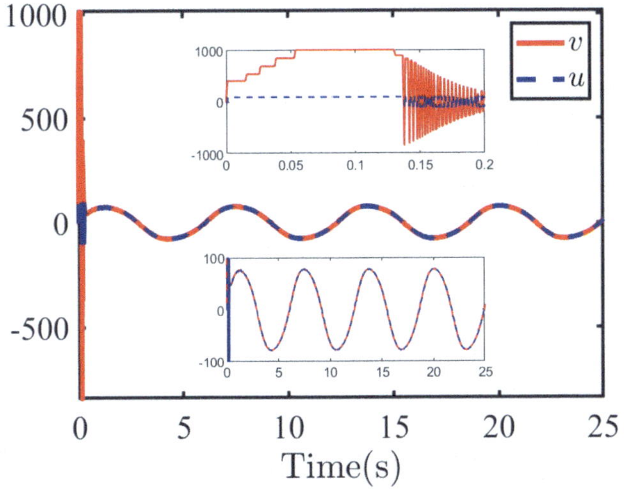

Fig. 4.12 Responses of v and u in Example 4.2

References

Chen M, Tao G, Jiang B (2015) Dynamic surface control using neural networks for a class of uncertain nonlinear systems with input saturation. IEEE Trans Neural Netw Learn Syst 26(9):2086–2097

Chen Z, Han QL, Yan Y, Wu ZG (2020) How often should one update control and estimation: review of networked triggering techniques. Sci China Inf Sci 63:150201

Chen J, Li J, Yuan X (2020) Global fuzzy adaptive consensus control of unknown nonlinear multiagent systems. IEEE Trans Fuzzy Syst 28(3):510–522
Chen M, Xiong S, Wu Q (2021) Tracking flight control of quadrotor based on disturbance observer. IEEE Trans Syst Man Cybernet Syst 51(3):1414–1423
Chen J, Li J, Guo Y, Li J (2022) Consensus control of mixed-order nonlinear multiagent systems: framework and case study. IEEE Trans Cybernet 52(12):13073–13082
Cui R, Yang C, Li Y, Sharma S (2017) Adaptive neural network control of AUVs with control input nonlinearities using reinforcement learning. IEEE Trans Syst Man Cybernet Syst 47(6):1019–1029
Ding D, Wang Z, Ho DW, Wei G (2017) Observer-based event-triggering consensus control for multiagent systems with lossy sensors and cyber-attacks. IEEE Trans Cybernet 47(8):1936–1947
Gao H, Li Z, Yu X, Qiu J (2022) Hierarchical multiobjective heuristic for PCB assembly optimization in a beam-head surface mounter. IEEE Trans Cybernet 52(7):6911–6924
Ge SS, Wang C, Lee TH (2000) Adaptive backstepping control of a class of chaotic systems. Int J Bifurcat Chaos 10(05):1149–1156
Guo J, Luo Y, Wang J, Li K, Chen T (2019) An adaptive cascade trajectory tracking control for over-actuated autonomous electric vehicles with input saturation. Sci China Technol Sci 62(12):2153–2160
Han H, Zhang J, Du S, Sun H, Qiao J (2021) Robust optimal control for anaerobic-anoxic-oxic reactors. Sci China Technol Sci 64(7):1485–1499
He W, Dong Y, Sun C (2016) Adaptive neural impedance control of a robotic manipulator with input saturation. IEEE Trans Syst Man Cybernet Syst 46(3):334–344
He X, Wang Q, Hao Y (2020) Finite-time adaptive formation control for multi-agent systems with uncertainties under collision avoidance and connectivity maintenance. Sci China Technol Sci 63(11):2305–2314
Heemels WH, Donkers M, Teel AR (2013) Periodic event-triggered control for linear systems. IEEE Trans Autom Control 58(4):847–861
Hu A, Park J, Cao J, Hu M, Luo Y (2021) Event-triggered bipartite consensus over cooperation-competition networks under DoS attacks. Sci China Technol Sci 64(1):157–168
Hu X, Li YX, Hou Z, Niu B (2021) Event-triggered prescribed performance adaptive fuzzy asymptotic tracking of nonstrict-feedback nonlinear systems. Int J Robust Nonlinear Control 31(12):5776–5795
Hu H, Wang B, Cheng Z, Liu L, Wang Y, Luo X (2021) A novel active fault-tolerant control for spacecrafts with full state constraints and input saturation. Aerosp Sci Technol 108:106368
Huang C, Cao J (2019) Comparative study on bifurcation control methods in a fractional-order delayed predator-prey system. Sci China Technol Sci 62:298–307
Huang B, Song S, Zhu C, Li J, Zhou B (2021) Finite-time distributed formation control for multiple unmanned surface vehicles with input saturation. Ocean Eng 233:109158
Jia T, Pan Y, Liang H, Lam HK (2022) Event-based adaptive fixed-time fuzzy control for active vehicle suspension systems with time-varying displacement constraint. IEEE Trans Fuzzy Syst 30(8):2813–2821
Lei Y, Li J, Zhao A (2022) Adaptive neural networks control for uncertain parabolic distributed parameter systems with nonlinear periodic time-varying parameter. Sci China Technol Sci 65(7):1482–1492
Li Z, Huang B, Ye Z, Deng M, Yang C (2018) Physical human-robot interaction of a robotic exoskeleton by admittance control. IEEE Trans Indus Electron 65(12):9614–9624
Li Z, Ren Z, Zhao K, Deng C, Feng Y (2020) Human-cooperative control design of a walking exoskeleton for body weight support. IEEE Trans Indus Inf 16(5):2985–2996
Li Z, Deng C, Zhao K (2020) Human-cooperative control of a wearable walking exoskeleton for enhancing climbing stair activities. IEEE Trans Indus Electron 67(4):3086–3095
Li H, Wu Y, Chen M (2021) Adaptive fault-tolerant tracking control for discrete-time multiagent systems via reinforcement learning algorithm. IEEE Trans Cybernet 51(3):1163–1174

Li Z, Zhao K, Zhang L, Wu X, Zhang T, Li Q, Li X, Su CY (2021) Human-in-the-loop control of a wearable lower limb exoskeleton for stable dynamic walking. IEEE/ASME Trans Mechatron 26(5):2700–2711

Li Z, Yu X, Qiu J, Gao H (2022) Cell division genetic algorithm for component allocation optimization in multifunctional placers. IEEE Trans Indus Inf 18(1):559–570

Liang H, Liu G, Zhang H, Huang T (2021) Neural-network-based event-triggered adaptive control of nonaffine nonlinear multiagent systems with dynamic uncertainties. IEEE Trans Neural Netw Learn Syst 32(5):2239–2250

Liang H, Guo X, Pan Y, Huang T (2021) Event-triggered fuzzy bipartite tracking control for network systems based on distributed reduced-order observers. IEEE Trans Fuzzy Syst 29(6):1601–1614

Lin G, Li H, Ma H, Yao D, Lu R (2022) Human-in-the-loop consensus control for nonlinear multi-agent systems with actuator faults. IEEE/CAA J Automatica Sinica 9(1):111–122

Liu YJ, Zeng Q, Tong S, Chen CP, Liu L (2019) Adaptive neural network control for active suspension systems with time-varying vertical displacement and speed constraints. IEEE Trans Indus Electron 66(12):9458–9466

Liu X, Xia J, Wang J, Shen H (2021) Interval type-2 fuzzy passive filtering for nonlinear singularly perturbed PDT-switched systems and its application. J Syst Sci Complexity 34(6):2195–2218

Liu Z, Lin W, Yu X, Rodríguez-Andina JJ, Gao H (2022) Approximation-free robust synchronization control for dual-linear-motors-driven systems with uncertainties and disturbances. IEEE Trans Indus Electron 69(10):10500–10509

Ma H, Li H, Lu R, Huang T (2020) Adaptive event-triggered control for a class of nonlinear systems with periodic disturbances. Sci China Inf Sci 63:150212

Na J, Huang Y, Wu X, Su SF, Li G (2020) Adaptive finite-time fuzzy control of nonlinear active suspension systems with input delay. IEEE Trans Cybernet 50(6):2639–2650

Pan Y, Du P, Xue H, Lam HK (2021) Singularity-free fixed-time fuzzy control for robotic systems with user-defined performance. IEEE Trans Fuzzy Syst 29(8):2388–2398

Pan Y, Wu Y, Lam HK (2022) Security-based fuzzy control for nonlinear networked control systems with DoS attacks via a resilient event-triggered scheme. IEEE Trans Fuzzy Syst 30(10):4359–4368

Pan Y, Li Q, Liang H, Lam HK (2022) A novel mixed control approach for fuzzy systems via membership functions online learning policy. IEEE Trans Fuzzy Syst 30(9):3812–3822

Qiu J, Sun K, Rudas IJ, Gao H (2020) Command filter-based adaptive NN control for MIMO nonlinear systems with full-state constraints and actuator hysteresis. IEEE Trans Cybernet 50(7):2905–2915

Shen J, Lam J (2022) Analysis of positive systems with input saturation: invariant hyperpyramids and hyperrectangles. IEEE Trans Autom Control 67(6):3005–3012

Tong S, Min X, Li Y (2020) Observer-based adaptive fuzzy tracking control for strict-feedback nonlinear systems with unknown control gain functions. IEEE Trans Cybernet 50(9):3903–3913

Wang L, Chen CP (2021) Reduced-order observer-based dynamic event-triggered adaptive NN control for stochastic nonlinear systems subject to unknown input saturation. IEEE Trans Neural Netw Learn Syst 32(4):1678–1690

Wang X, Liu Z, Chen Z (2021) Event-triggered fault-tolerant consensus control with control allocation in leader-following multi-agent systems. Sci China Technol Sci 64(4):879–889

Wang Q, Zhuang H, Duan Z, Wang Q (2021) Robust control of uncertain robotic systems: an adaptive friction compensation approach. Sci China Technol Sci 64(6):1228–1237

Wang J, Yang C, Xia J, Wu ZG, Shen H (2022) Observer-based sliding mode control for networked fuzzy singularly perturbed systems under weighted try-once-discard protocol. IEEE Trans Fuzzy Syst 30(6):1889–1899

Wu K, Zhang P, Wu H (2019) A new control design for a morphing UAV based on disturbance observer and command filtered backstepping techniques. Sci China Technol Sci 62:1845–1853

Xia X, Cheng L (2021) Adaptive Takagi-Sugeno fuzzy model and model predictive control of pneumatic artificial muscles. Sci China Technol Sci 64(10):2272–2280

Xu Y, Wu ZG (2021) Distributed adaptive event-triggered fault-tolerant synchronization for multiagent systems. IEEE Trans Indus Electron 68(2):1537–1547

Yang C, Jiang Y, He W, Na J, Li Z, Xu B (2018) Adaptive parameter estimation and control design for robot manipulators with finite-time convergence. IEEE Trans Indus Electron 65(10):8112–8123

Zhang Y, Li H, Sun J, He W (2019) Cooperative adaptive event-triggered control for multiagent systems with actuator failures. IEEE Trans Syst Man Cybernet Syst 49(9):1759–1768

Zhang H, Liu Y, Wang Y (2021) Observer-based finite-time adaptive fuzzy control for nontriangular nonlinear systems with full-state constraints. IEEE Trans Cybernet 51(3):1110–1120

Zhang H, Xi R, Wang Y, Sun S, Sun J (2022) Event-triggered adaptive tracking control for random systems with coexisting parametric uncertainties and severe nonlinearities. IEEE Trans Autom Control 67(4):2011–2018

Zhao X, Yang H, Xia W, Wang X (2017) Adaptive fuzzy hierarchical sliding-mode control for a class of MIMO nonlinear time-delay systems with input saturation. IEEE Trans Fuzzy Syst 25(5):1062–1077

Chapter 5
Adaptive Fuzzy Control with Output Constraint

Abstract In this chapter, an adaptive fuzzy controller is designed for a single-link flexible-joint robotic manipulator subject to output constraint, in which the fuzzy logic systems are employed to identify the unknown nonlinearity. The barrier Lyapunov function is constructed to guarantee that the output of the single-link flexible-joint robotic manipulator is not to violate its constraint. In order to reduce the occupation of communication channel bandwidth, the event-triggered control law is developed based on fixed threshold approach. Incorporating the backstepping technique and Lyapunov stability theory, the proposed control strategy ensures all variables in the flexible-joint robotic manipulator are bounded, and the tracking error converges to the origin with a small scale. Simulation results are given to illustrate the efficacy of the developed control strategy.

5.1 Introduction

Due to the fact that the robotic manipulators are highly coupled, nonlinear, and have well defined dynamic models, considerable attention has been paid on the tracking control of robotic manipulators (Hayward and Paul 1986; Kallu et al. 2018; Xiao and Yin 2018). However, in most robotic applications, it is not possible to obtain all the precise system parameters related to the model. In addition, the presence of unstructured payloads, friction effects, and external interference make it difficult to obtain ideal control performance. Therefore, the fuzzy logic systems (FLSs) or neural networks were used to provide robustness to the parametric uncertainties of robotic manipulators (Li et al. 2020; Li and Li 2017; Huang et al. 2020; Zhou et al. 2019; Zhang et al. 2021; Xiao et al. 2020; Yoo and Ham 2000). To mention a few, for the robotic manipulators against state constraints and unknown system dynamics (He and Dong 2017), the fuzzy neural network learning algorithm was developed to identify the uncertain plant model. By using FLSs to identify unknown nonlinearities, an approximation-based adaptive fuzzy controller was designed for robot finger dynamics (Zhou et al. 2014).

Recently, barrier Lyapunov function (BLF) has been widely employed to handle the nonlinear systems control problem with output and state constraints. In the work

H. Ren et al., *Intelligent Control of Nonlinear Systems and Nonlinear Multi-Agent Systems with Complex Constraints*, Intelligent Control and Learning Systems 16,
https://doi.org/10.1007/978-981-96-0320-6_5

of Tee et al. (2009), symmetric and asymmetric BLFs were constructed to prevent constraint violation. The tracking controller was designed of a nonlinear switched lower triangular system (Niu et al. 2017), in which the switched constrained system was transformed into a novel system without any constraint by introducing a novel BLF firstly. On the other hand, event-triggered control strategies can reduce the communication resource occupancy. Due to the fact that the communication resource is limited, it is essential to reduce the communication burden (Li et al. 2018; Ma et al. 2020; Zhou et al. 2019; Zhang et al. 2020). For instance, an event-triggered scheduler has been investigated in Tabuada (2007), which guaranteed the input-to-state stability. In Heemels et al. (2012), the use of output-based event-triggered control has been provided, in which the differences between two event-triggering mechanisms have been stated. For the networked control plants, the dynamic output feedback control problem was discussed in Pan and Yang (2017) based on event-triggered communication mechanism. However, the fuzzy event-triggered control issue has not been fully investigated of a single-link flexible-joint robotic manipulator subject to output constraint.

Inspired by the aforementioned works, an approximation-based fuzzy adaptive event-triggered control approach is developed of a single-link flexible-joint robotic manipulator subject to output constraint in this chapter. By introducing the BLF, the constructed adaptive fuzzy event-triggered controller ensures that all variables of the single-link flexible-joint robot are semi-globally uniformly ultimately bounded (SGUUB) and the single-link flexible-joint robot output is not to violate its constraint. In addition, the simulation results show the feasibility of this approach.

5.2 Preliminaries and Problem Formulation

5.2.1 System Formulation

Figure 5.1 shows the single-link flexible-joint robotic manipulator structure, which can be described by

$$\begin{aligned} I\ddot{q}_1 + MgL\sin q_1 + K(q_1 - q_2) &= 0 \\ J\ddot{q}_2 - K(q_1 - q_2) &= u, \end{aligned} \tag{5.1}$$

where u denotes the input torque of the control system. I and J stand for the inertia of the link and motor, repsectively. K shows the spring constant of joint stiffness. M expresses the link mass. g denotes the gravity term and L indicates the distance from the centroid of the link to the axis of the joint. It should be noted that the angular position of the link q_1 and motor shaft q_2 are available.

Define $x_1 = q_1$, $x_2 = \dot{q}_1$, $x_3 = q_2$, $x_4 = \dot{q}_2$. Therefore, the single-link flexible-joint robot is expressed as follows:

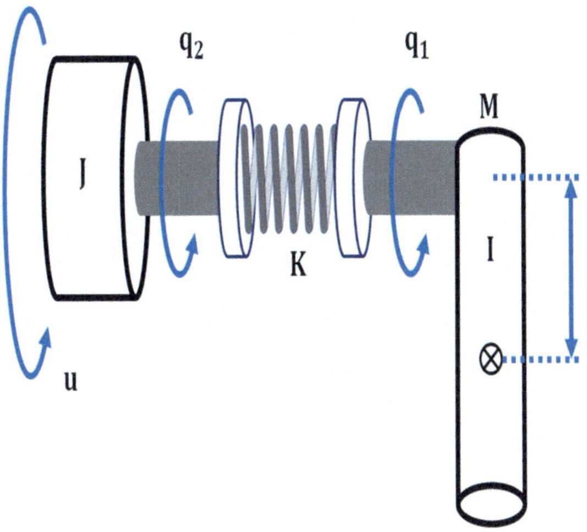

Fig. 5.1 Single-link flexible-joint robotic manipulator

$$\begin{aligned}\dot{x}_1 &= x_2\\ \dot{x}_2 &= -\frac{K}{I}(x_1 - x_3) - \frac{MgL}{I}\sin x_1\\ \dot{x}_3 &= x_4\\ \dot{x}_4 &= \frac{1}{J}u + \frac{K}{J}(x_1 - x_3)\\ y &= x_1,\end{aligned} \tag{5.2}$$

where u and y represent the system control input and system output, respectively. x_i denotes the control system state with $i = 1, 2, 3, 4$.

5.2.2 Preliminaries

Assumption 5.1 (Liu et al. 2018) The reference signal $y_d(t)$, and its ith-order derivatives $y_d^{(i)}(t)$ are sufficiently smooth and bounded. Supposing that $|y_d(t)| \leq \mathcal{A}_0 < k_{c1}$ and $\left|y_d^{(i)}(t)\right| < \mathcal{A}_i$ with $i = 1, 2, 3, 4$, where k_{c1} *and* $\mathcal{A}_i$ $(i = 0, 1, 3, 4)$ are positive constants.

Lemma 5.1 (Li and Li 2017) *There exist positive constants k_{ci} and variables z_1, which satisfies $\Omega_z = \{z \in R| \, |z_1| < k_{c1}\}$. The following formula holds*

$$\log \frac{k_{c1}^2}{k_{c1}^2 - z_1^2} < \frac{z_1^2}{k_{c1}^2 - z_1^2}$$

for the interval Ω_z.

Lemma 5.2 (Zhou et al. 2014) *For a positive constant ε, one has*

$$xy \le \frac{1}{q\varepsilon^q}|y|^q + \frac{\varepsilon^p}{p}|x|^p$$

where $(p-1)(q-1)=1$ with $p>1$ and $q>1$.

Lemma 5.3 (Xing et al. 2017) *For $\Upsilon \in R$, one get*

$$0 \le |\Upsilon| - \Upsilon \tanh\left(\frac{\Upsilon}{\epsilon}\right) \le 0.2785\epsilon,$$

where ϵ denotes a positive parameter.

Assumption 5.2 (Chen et al. 2015) There exists strictly increasing smooth function $\psi_i(\cdot): R^+ \to R^-$ with $\psi_i(0)=0$ such that

$$|f_i(x)| \le \psi_i(||x||).$$

Remark 5.1 For $a_j \ge 0$ $(j=1,2,\ldots,n)$, we can get $\psi_i(\sum_{j=1}^n a_j) \le \sum_{j=1}^n \psi_i(na_j)$ by the increasing property of $\psi_i(\cdot)$. Therefore, there exists a smooth function $h_i(na_j)$ such that $\psi_i(na_j) = na_j h_i(na_j)$, which results in

$$\psi_i\left(\sum_{j=1}^n a_j\right) \le \sum_{j=1}^n na_j h_i(na_j).$$

Lemma 5.4 (Chen et al. 2015) *For $z_i = x_i - \alpha_{i-1}$ and $i=1,2,3,4$, one has*

$$||x|| \le \sum_{i=1}^4 |z_i|\phi_i(\cdot) + \mathcal{A}_0$$

with

$$\begin{aligned}
\phi_1(\cdot) &= 1 + m_1 + \frac{1+\frac{b_1}{\zeta_1^2}}{2(k_{c1}^2 - z_1^2)} + \frac{b_1\hat{\theta}_1\Phi_1^T\Phi_1}{2\xi_1^2(k_{c1}^2 - z_1^2)} \\
\phi_i(\cdot) &= 1 + m_i + \frac{b_i}{2\zeta_i^2} + \frac{b_i}{2\xi_i^2}\hat{\theta}_i\Phi_i^T\Phi_i,\ i=2,3 \\
\phi_4(\cdot) &= 1,
\end{aligned}$$

where m_i, b_i, ζ_i and ξ_i $(i=1,2,3,4)$ are positive parameters and $\alpha_0 = y_d$.

5.2.3 FLSs

In order to identify the unknown nonlinearities in the single-link flexible-joint robot, FLSs are adopted in the control designing, and it is described as
Rule j: IF x_1 is $\mathcal{M}_1^j$, ..., and x_n is $\mathcal{M}_n^j$, THEN

$$y\ is\ N_j,$$

where $j = 1, \ldots, N$, y and $x = [x_1, \ldots, x_n]^T$ denote output and input of FLSs, respectively. From Gandhi and Adhyaru (2020), the FLSs are described as

$$y(x) = \frac{\sum\limits_{j=1}^{N} \tilde{y}_j \prod\limits_{i=1}^{n} \gamma_{\mathcal{M}_i^j}(x_i)}{\sum\limits_{j=1}^{N} \prod\limits_{i=1}^{n} \gamma_{\mathcal{M}_i^j}(x_i)},$$

where $\tilde{y}_j = \max_{y \in R} \gamma_{N_j}(y)$. The fuzzy sets are $\mathcal{M}_i^j$ and N_j. The membership functions are represented by $\gamma_{\mathcal{M}_i^j}(x_i)$ and $\gamma_{N_j}(y)$, respectively. For $j = 1, 2, \ldots, N$, the fuzzy basis functions are

$$\Phi_j(x) = \frac{\prod_{i=1}^{n} \gamma_{\mathcal{M}_i^j}(x_i)}{\sum_{j=1}^{N} \prod_{i=1}^{n} \gamma_{\mathcal{M}_i^j}(x_i)}.$$

Then, one gets

$$y(x) = W^T \Phi(x),$$

where

$$\begin{aligned} \Phi(x) &= [\Phi_1(x), \Phi_2(x), \ldots, \Phi_N(x)]^T \\ W^T &= [\tilde{y}_1, \tilde{y}_2, \ldots, \tilde{y}_N] = [W_1, W_2, \ldots, W_N]. \end{aligned}$$

Lemma 5.5 (Wang 1993) *For a continuous bounded function $F(x)$ which defined on a compact Ω, then there exists a FLS such that the following formula holds*

$$\sup_{x \in \Omega} \left| F(x) - W^T \Phi(x) \right| \le \mu$$

for any scalar $\mu > 0$.

5.3 Event-Triggered Control Strategy Design and Stability Analysis

In this chapter, the symmetric BLF is used to guarantee that the output of the robotic manipulator is not to violate its constraint. Define

$$\begin{aligned} z_1 &= x_1 - y_d \\ z_i &= x_i - \alpha_{i-1},\ i = 2, 3, 4, \end{aligned}$$

where y_d refers to the desired reference signal. α_{i-1} $(i = 2, 3, 4)$ represents the virtual controllers, and it can be expressed as

$$\begin{aligned} \alpha_1 &= -m_1 z_1 - \frac{z_1}{2(k_{c1}^2 - z_1^2)} - \frac{b_1 z_1 \hat{\theta}_1 \Phi_1^T \Phi_1}{2\xi_1^2 (k_{c1}^2 - z_1^2)} \\ &\quad - \frac{b_1 z_1}{2\zeta_1^2 (k_{c1}^2 - z_1^2)} \\ \alpha_i &= -m_i z_i - \frac{b_i z_i}{2\xi_i^2} \hat{\theta}_i \Phi_i^T \Phi_i - \frac{b_i z_i}{2\zeta_i^2},\ i = 2, 3, 4, \end{aligned} \tag{5.3}$$

where m_i, b_i, ξ_i, and ζ_i $(i = 1, 2, 3, 4)$ are designed positive parameters. Followed by the virtual controllers, the adaptive laws are obtained as

$$\begin{aligned} \dot{\hat{\theta}}_1 &= r_1 \left(\frac{b_1 z_1^2 \Phi_1^T \Phi_1}{2\xi_1^2 (k_{c1}^2 - z_1^2)^2} - \sigma_1 \hat{\theta}_1 \right) \\ \dot{\hat{\theta}}_i &= r_i \left(\frac{b_i z_i^2}{2\xi_i^2} \Phi_i^T \Phi_i - \sigma_i \hat{\theta}_i \right),\ i = 2, 3, 4, \end{aligned} \tag{5.4}$$

where $\hat{\theta}_i$ is the estimation of θ_i, $\tilde{\theta}_i = \theta_i - \hat{\theta}_i$. r_i and σ_i are positive parameters to be designed. By using the fixed threshold method (Xing et al. 2016), the adaptive event-triggered controller can be designed as

$$w(t) = -\eta \tanh\left(\frac{z_4 \eta}{\epsilon}\right) + \alpha_4. \tag{5.5}$$

Define the event-triggering mechanism

$$\begin{aligned} t_{k+1} &= \inf\{t > t_k \mid |e(t)| \geq \bar{\eta}\},\quad t_1 = 0 \\ u(t) &= w(t_k),\ \forall t \in [t_k, t_{k+1}), \end{aligned}$$

where $\varepsilon > 0$, $\eta > \bar{\eta} > 0$, and $e(t) = w(t) - u(t)$. Moreover, it should be noted that the designed controller in this chapter can avoid Zeno behavior. Proof please refer to the work in Xing et al. (2016).

Choosing the positive BLF as

$$V=\frac{1}{2}\left(\log\frac{k_{c1}^2}{k_{c1}^2-z_1^2}+\sum_{i=2}^{3}z_i^2+Jz_4^2+\sum_{i=1}^{4}\frac{1}{r_i}\tilde{\theta}_i^T\tilde{\theta}_i\right).$$

From the system (5.2), we have

$$\begin{aligned}\dot{V}=&\frac{z_1}{k_{c1}^2-z_1^2}(x_2-\dot{y}_d)+z_2(x_3+\bar{f}(x_1,x_3)-\dot{\alpha}_1)\\&+z_3(x_4-\dot{\alpha}_2)+z_4[u+k(x_1-x_3)-\dot{\alpha}_3]-\sum_{i=1}^{4}\frac{1}{r_i}\tilde{\theta}_i^T\dot{\hat{\theta}}_i\end{aligned}$$

with $\bar{f}(x_1,x_3)=-x_3-\frac{K}{I}(x_1-x_3)-\frac{MgL}{I}\sin x_1$. From Assumption 5.2, Remark 5.1 and Lemma 5.4, one gets

$$z_2\bar{f}(x_1,x_3)\le\frac{25}{2}\sum_{i=1}^{4}z_i^2\phi_i^2(\cdot)h^2(5|z_i|\phi_i(\cdot))+|z_2|\psi(5\mathcal{A}_0)+2z_2^2.$$

Moreover, based on Lemma 5.2, the following expression holds

$$|z_2|\psi(5\mathcal{A}_0)\le z_2\psi(5\mathcal{A}_0)\tanh\left(\frac{z_2\psi(5\mathcal{A}_0)}{\epsilon}\right)+0.2785\epsilon.$$

From Young's inequality, one gets that the following Inequalities:

$$\begin{aligned}\frac{z_1z_2}{k_{c1}^2-z_1^2}\le&\frac{1}{2}\left(\frac{z_1}{k_{c1}^2-z_1^2}\right)^2+\frac{1}{2}z_2^2\\z_iz_{i+1}\le&\frac{1}{2}z_i^2+\frac{1}{2}z_{i+1}^2,\quad i=2,3.\end{aligned}$$

The nonlinear functions F_i $(i=1,2,3,4)$ are defined as

$$\begin{aligned}F_1=&\frac{25}{2}(k_{c1}^2-z_1^2)|z_1|\phi_1^2(\cdot)h^2(5|z_1|\phi_1(\cdot))-\dot{y}_d\\F_2=&3z_2+\frac{25}{2}|z_2|\phi_2^2(\cdot)h^2(5|z_2|\phi_2(\cdot))\\&+\psi(5\mathcal{A}_0)\tanh\left(\frac{z_2\psi(5\mathcal{A}_0)}{\epsilon}\right)-\frac{\partial\alpha_1}{\partial x_1}x_2\\&-\frac{\partial\alpha_1}{\partial\hat{\theta}_1}\dot{\hat{\theta}}_1-\frac{\partial\alpha_1}{\partial y_d}\dot{y}_d\end{aligned}$$

$$
\begin{aligned}
F_3 = &\ z_3 + \frac{25}{2}|z_3|\phi_3^2(\cdot)h^2(5|z_3|\phi_3(\cdot)) - \sum_{i=1}^{2}\frac{\partial \alpha_2}{\partial x_i}\dot{x}_i \\
&- \sum_{i=1}^{2}\frac{\partial \alpha_2}{\partial \hat{\theta}_i}\dot{\hat{\theta}}_i - \sum_{i=1}^{2}\frac{\partial \alpha_2}{\partial y_d^{(i-1)}}y_d^{(i)} \\
F_3 = &\ \frac{1}{2}z_4 + k(x_1 - x_3) + \frac{25}{2}|z_4|\phi_4^2(\cdot)h^2(5|z_4|\phi_4(\cdot)) \\
&- \sum_{i=1}^{3}\frac{\partial \alpha_3}{\partial x_i}\dot{x}_i - \sum_{i=1}^{3}\frac{\partial \alpha_3}{\partial \hat{\theta}_i}\dot{\hat{\theta}}_i - \sum_{i=1}^{3}\frac{\partial \alpha_3}{\partial y_d^{(i-1)}}y_d^{(i)}.
\end{aligned}
$$

The FLSs are used to approximate F_i $(i = 1, 2, 3, 4)$, which are described by

$$
F_i = W_i^T\Phi_i + \mu_i,\ |\mu_i| \le \varepsilon_i.
$$

Then, according to the Lemma 5.2, one has

$$
\begin{aligned}
\frac{z_1}{k_{c1}^2 - z_1^2}F_1 \le &\ \frac{b_1 z_1^2}{2\xi_1^2(k_{c1}^2 - z_1^2)}\theta_1\Phi_1^T\Phi_1 + \frac{\xi_1^2}{2} \\
&+ \frac{b_1}{2\zeta_1^2(k_{c1}^2 - z_1^2)}z_1^2 + \frac{\zeta_1^2}{2b_1}\varepsilon_1^2 \\
z_i F_i \le &\ \frac{b_i z_i^2}{2\xi_i^2}\theta_i\Phi_i^T\Phi_i + \frac{\xi_i^2}{2} \\
&+ \frac{b_i}{2\zeta_i^2}z_i^2 + \frac{\zeta_i^2}{2b_i}\varepsilon_i^2,\ i = 2, 3, 4,
\end{aligned}
$$

where $\theta_i = \frac{||W_i||^2}{b_i}$ with $i = 1, 2, 3, 4$. Based on the Lemma 5.3, coordinate transformation, virtual controllers, adaptive laws, and the above inequalities, one has

$$
\begin{aligned}
\dot{V} \le &\ -\frac{m_1 z_1^2}{k_{c1}^2 - z_1^2} - \sum_{i=2}^{4} m_i z_i^2 - \sum_{i=1}^{4}\frac{1}{2}\sigma_i\tilde{\theta}_i^2 \\
&+ \sum_{i=1}^{4}\left(\frac{\xi_i^2}{2} + \frac{\zeta_i^2}{2b_i}\varepsilon_i^2 + \frac{1}{2}\sigma_i\theta_i^2\right) + 0.557\epsilon.
\end{aligned}
$$

Finally, the following expression holds

$$
\dot{V} \le -a_0 V + b_0,
$$

where $a_0 = \min\{2m_1, 2m_2, 2m_3, 2m_4/J, \sigma_i\}$ and

$$b_0 = 0.557\epsilon + \sum_{i=1}^{4}\left(\frac{\xi_i^2}{2} + \frac{\zeta_i^2}{2b_i}\varepsilon_i^2 + \frac{1}{2}\sigma_i\theta_i^2\right), \quad i = 1, 2, 3, 4.$$

Multiplying both sides by $e^{a_0 t}$, integrating it over $[0, t]$, and based on Lemma 5.1, the following inequality holds

$$\log\frac{k_{c1}^2}{k_{c1}^2 - z_1^2} \leq 2\left[V(0) - \frac{b_0}{a_0}\right]e^{-a_0 t} + \frac{2b_0}{a_0}.$$

Then, we have

$$|z_1| \leq k_{c1}\sqrt{1 - e^{-\frac{2b_0}{a_0}}}.$$

Therefore, it can be seen that all the variables in the single-link flexible-joint robotic manipulator are SGUUB, and by designing adequate parameters, the output can track the reference signal. These results can be summarized in Theorem 5.1.

Theorem 5.1 *For the single-link flexible-joint robotic manipulator* (5.1) *under Assumptions 5.1–5.2, event-triggering rules, the virtual control signals* (5.3), *the adaptive laws* (5.4), *the intermediate control law* (5.5), *then, it is proved that all signals in the single-link flexible-joint robotic manipulator are SGUUB and the tracking error is remained in a bounded compact set.*

5.4 Simulation Results

Simulation results are rendered to show the efficacy of the developed control strategy in this section.

The single-link flexible-joint robotic manipulator parameters are chosen as given in Zhang et al. (2010). The parameters are $r_i = \xi_i = \zeta_i = 1$ $(i = 1, 2, 3, 4)$, $\sigma_1 = 1$, $m_1 = 10$, $m_2 = m_3 = m_4 = 5$, $k_{c1} = 5.5$, $\eta = 2$. Simulation results are described in Figs. 5.2, 5.3, 5.4, 5.5, 5.6, and 5.7. In Fig. 5.2, we can see the position state q_1 can primely tracks y_d. with output constraint. The responses of x_2, x_3 and x_4 are shown in Figs. 5.3, 5.4 and 5.5. For the fixed threshold strategy, Fig. 5.6 shows the time intervals of each event. Moreover, Fig. 5.7 presents the response of control input u.

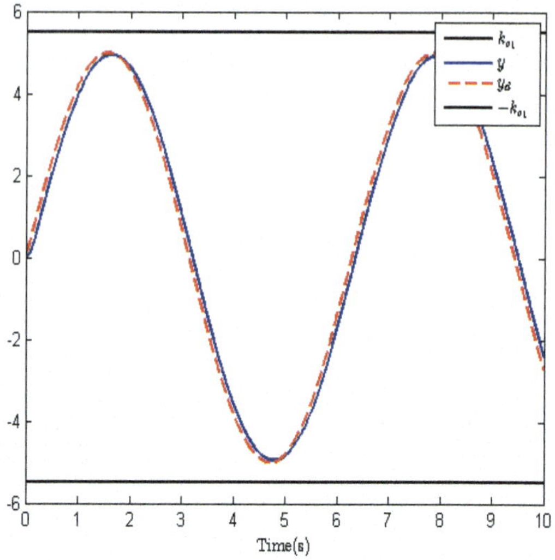

Fig. 5.2 Trajectories of y and the reference signal y_d

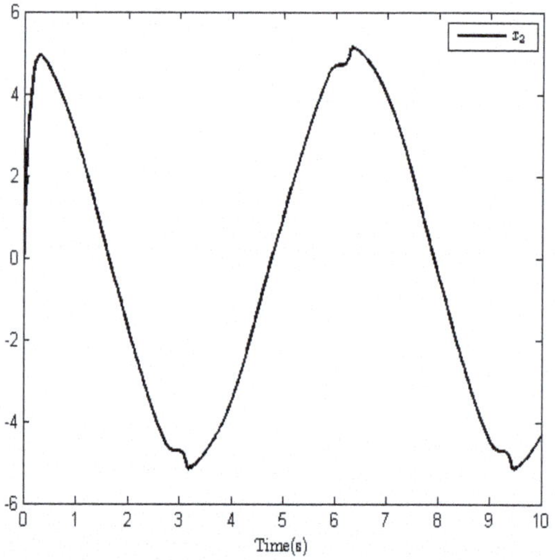

Fig. 5.3 Trajectory of system states x_2

5.5 Conclusion

For a flexible-joint robot subject to output constraint, the fuzzy event-triggered control strategy has been developed. It is pointed that a reliable adaptive event-triggered controller, which is based on a fixed threshold strategy, has been designed to reduce the communication channel bandwidth. Moreover, the obstacle of output constraint

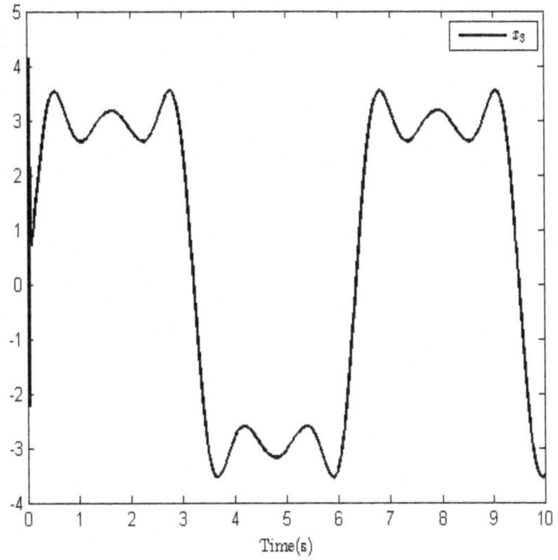

Fig. 5.4 Trajectory of system states x_3

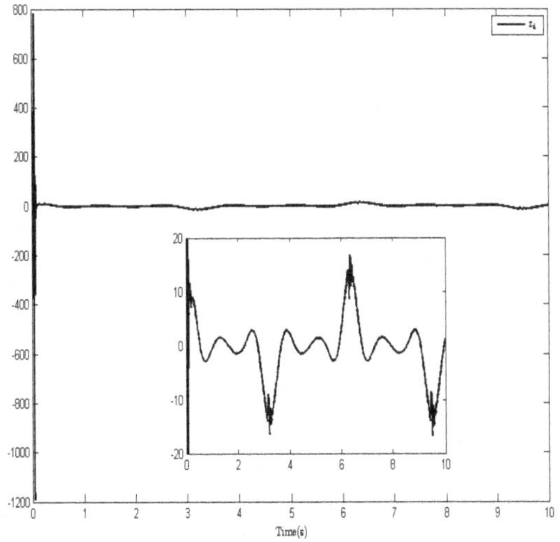

Fig. 5.5 Trajectory of system states x_4

has been solved based on a symmetric BLF. Incorporating the Lyapunov stability theory and backstepping technology, an adaptive fuzzy event-triggered controller has been designed to ensure that all variables of the closed-loop system are bounded, and the reference signal can be followed by the system output.

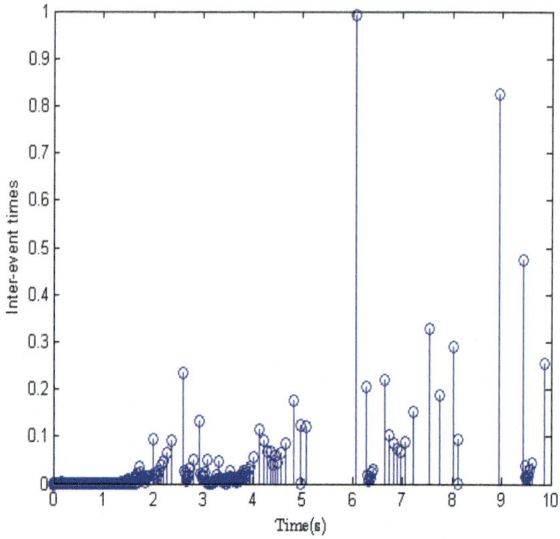

Fig. 5.6 Time interval of triggering events

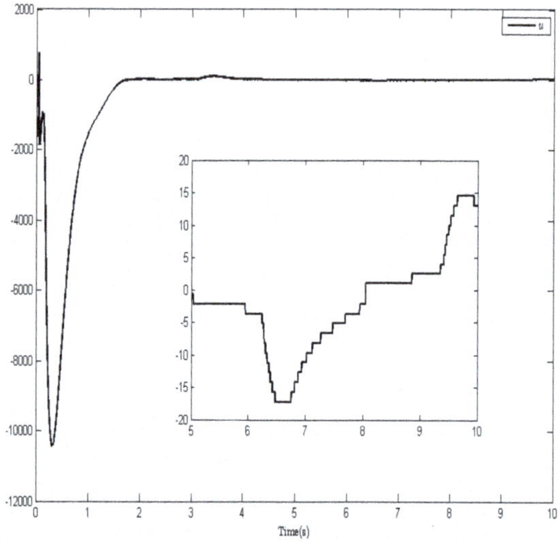

Fig. 5.7 Trajectory of control input

References

Chen B, Zhang H, Lin C (2015) Observer-based adaptive neural network control for nonlinear systems in nonstrict-feedback form. IEEE Trans Neural Netw Learn Syst 27(1):89–98

Gandhi RV, Adhyaru DM (2020) Takagi-Sugeno fuzzy regulator design for nonlinear and unstable systems using negative absolute eigenvalue approach. IEEE/CAA J Automatica Sinica 7(2):482–493

Hayward V, Paul RP (1986) Robot manipulator control under Unix RCCL: a robot control "C" library. Int J Robot Res 5(4):94–111

He W, Dong Y (2017) Adaptive fuzzy neural network control for a constrained robot using impedance learning. IEEE Trans Neural Netw Learn Syst 29(4):1174–1186
Heemels WPMH, Johansson KH, Tabuada P (2012) An introduction to event-triggered and self-triggered control. Decision Control 3270–3285
Huang H, Zhang T, Yang C, Chen CP (2020) Motor learning and generalization using broad learning adaptive neural control. IEEE Trans Indus Electron 67(10):8608–8617
Kallu KD, Jie W, Lee MC (2018) Sensorless reaction force estimation of the end effector of a dual-arm robot manipulator using sliding mode control with a sliding perturbation observer. Int J Control Autom Syst 16(3):1367–1378
Li DP, Li DJ (2017) Adaptive neural tracking control for an uncertain state constrained robotic manipulator with unknown time-varying delays. IEEE Trans Syst Man Cybernet Syst 48(12):2219–2228
Li H, Zhang Z, Yan H, Xie X (2018) Adaptive event-triggered fuzzy control for uncertain active suspension systems. IEEE Trans Cybernet 49(12):4388–4397
Li Z, Gao L, Chen W, Xu Y (2020) Distributed adaptive cooperative tracking of uncertain nonlinear fractional-order multi-agent systems. IEEE/CAA J Automatica Sinica 7(1):292–300
Liu Y, Lu S, Tong S, Chen X, Chen CP, Li D (2018) Adaptive control-based barrier Lyapunov functions for a class of stochastic nonlinear systems with full state constraints. Automatica 87:83–93
Ma H, Li H, Lu R, Huang T (2020) Adaptive event-triggered control for a class of nonlinear systems with periodic disturbances. Sci China Inf Sci 63(5):150212
Niu B, Liu Y, Zong G, Han Z, Fu J (2017) Command filter-based adaptive neural tracking controller design for uncertain switched nonlinear output-constrained systems. IEEE Trans Cybernet 47(10):3160–3171
Pan Y, Yang G (2017) Event-triggered fuzzy control for nonlinear networked control systems. Fuzzy Sets Syst 329(15):91–107
Tabuada P (2007) Event-triggered real-time scheduling of stabilizing control tasks. IEEE Trans Autom Control 52(9):1680–1685
Tee KP, Ge SS, Tay EH (2009) Barrier Lyapunov functions for the control of output-constrained nonlinear systems. Automatica 45(4):918–927
Wang L (1993) Stable adaptive fuzzy control of nonlinear systems. IEEE Trans Fuzzy Syst 1(2):146–155
Xiao B, Yin S (2018) Exponential tracking control of robotic manipulators with uncertain dynamics and kinematics. IEEE Trans Indus Inf 15(2):689–698
Xiao W, Cao L, Li H, Lu R (2020) Observer-based adaptive consensus control for nonlinear multi-agent systems with time-delay. Sci China Inf Sci 63(3):132202
Xing L, Wen C, Liu Z, Su H, Cai J (2016) Event-triggered adaptive control for a class of uncertain nonlinear systems. IEEE Trans Autom Control 62(4):2071–2076
Xing L, Wen C, Liu Z, Su H, Cai J (2017) Adaptive compensation for actuator failures with event-triggered input. Automatica 85:129–136
Yoo BK, Ham WC (2000) Adaptive control of robot manipulator using fuzzy compensator. IEEE Trans Fuzzy Syst 8(2):186–199
Zhang X, Polycarpou MM, Parisini T (2010) Fault diagnosis of a class of nonlinear uncertain systems with Lipschitz nonlinearities using adaptive estimation. Automatica 46(2):290–299
Zhang XM, Han QL, Ge X, Ding D, Ding L, Yue D, Peng C (2020) Networked control systems: a survey of trends and techniques. IEEE/CAA J Automatica Sinica 7(1):1–17
Zhang T, Su G, Qing C, Xu X, Cai B, Xing X (2021) Hierarchical lifelong learning by sharing representations and integrating hypothesis. IEEE Trans Syst Man Cybernet Syst 51(2):1004–1014
Zhou Q, Li H, Shi P (2014) Decentralized adaptive fuzzy tracking control for robot finger dynamics. IEEE Trans Fuzzy Syst 23(3):501–510
Zhou Q, Chen G, Lu R, Bai W (2019) Disturbance-observer-based event-triggered control for multi-agent systems with input saturation. Scientia Sinica Informationis 49(11):1502–1516
Zhou Q, Zhao S, Li H, Lu R, Wu C (2019) Adaptive neural network tracking control for robotic manipulators with dead zone. IEEE Trans Neural Netw Learn Syst 30(12):3611–3620

Part II
Intelligent Control of Nonlinear MASs

Chapter 6
Adaptive Fixed-Time Control with Actuator Faults

Abstract The adaptive fixed-time consensus problem for a class of nonlinear multi-agent systems (MASs) with actuator faults is considered in this chapter. To approximate the unknown nonlinear functions in MASs, radial basis function neural networks are used. In addition, the first-order sliding mode differentiator is utilized to solve the "explosion of complexity" problem, and a filter error compensation method is proposed to ensure the convergence of filter error in fixed time. With the help of the Nussbaum function, the actuator failure compensation mechanism is constructed. By designing the adaptive fixed-time controller, all signals in MASs are bounded, and the consensus errors between the leader and all followers converge to a small area of origin. Finally, the effectiveness of the proposed control method is verified by simulation examples.

6.1 Introduction

The consensus control of multi-agent systems (MASs) has gained much attention in various fields in recent decades (Li et al. 2020; Deng et al. 2020; Wei et al. 2021; Liu et al. 2022; Zhang and Chai 2022; Lin et al. 2022; Ren et al. 2022; Liu et al. 2022; Zhang et al. 2023; Li et al. 2023; Jin et al. 2022, 2023, 2022). For consensus control, all followers in MASs should track the trajectory of leader based on messages from their local neighboring agents. The consensus control problem of high-order MASs is challenging. To solve this problem, the backstepping method was used for high-order MASs (Ni et al. 2021). In addition, neural networks (NNs) and fuzzy logic systems (FLSs) are usually utilized to deal with nonlinear functions in MASs (Zheng et al. 2023; Pan et al. 2022; Ma et al. 2022; Cao et al. 2023; Pan et al. 2022; Ren et al. 2023). For example, by combining the backstepping technique and FLS, the adaptive finite-time consensus controllers were obtained for high-order nonlinear MASs with unknown nonlinear dynamics in Zhou et al. (2022). Although backstepping is an effective method for high-order MASs, it will bring the "explosion of complexity" problem owing to repeated derivation of virtual control signals.

The dynamic surface control (DSC) method can be used to deal with the "explosion of complexity" problem. By constructing the first-order filter and using a virtual

H. Ren et al., *Intelligent Control of Nonlinear Systems and Nonlinear Multi-Agent Systems with Complex Constraints*, Intelligent Control and Learning Systems 16,
https://doi.org/10.1007/978-981-96-0320-6_6

signal as the input signal, the output of filter can be used to approximate the derivative of the virtual signal. For example, by combining the DSC approach and minimal learning parameter technique, the adaptive control method was proposed for nonlinear systems (Liu et al. 2016), in which the "explosion of complexity" problem was avoided. Although the work in Liu et al. (2016) can solve the above problem, the initial values of virtual signals were needed in these methods which are hard to obtain in practical systems. To remove the restrictions, the first-order sliding mode differentiator was used, which ensured that the filtered tracking errors were bounded, and the initial values were not needed (Chen and Ge 2015; Wang and Chen 2020). Based on the sliding mode differentiator, the tedious analytic computation problem was avoided in Chen and Ge (2015), and the state observer and disturbance observer were designed simultaneously to estimate unmeasured states and compounded disturbances. In Wang and Chen (2020), the sliding mode differentiator was utilized in the DSC strategy such that the "explosion of complexity" problem was handled, and the difficulties caused by unknown control directions and state constraints were well addressed.

The above results can only make all error signals converge to steady states when time goes to infinity. With a higher convergence rate, the finite-time control was studied in many practical applications (Dai et al. 2022; Li et al. 2020; Wang et al. 2020; Dai et al. 2020; Liu et al. 2022). Based on the finite-time control theory, the finite-time containment control method of nonlinear MASs was proposed in Li et al. (2021). A fuzzy finite-time command filtering output feedback control method was presented for a class of nonlinear systems (Wang et al. 2022). However, the convergence time of finite-time control depends on the initial condition of the system, which means that the convergence time may be unbounded if the initial condition is too far from the steady states. To overcome such a problem, the stability of fixed-time convergence was first considered in Polyakov et al. (2015). Different from the finite-time control strategy, the fixed-time control method ensures that the settling time is bounded by a constant, which avoids the influence caused by initial conditions. In Liu et al. (2019), the fixed-time consensus of nonlinear MASs was achieved, and the authors proved that the convergence time of fixed-time results is independent of initial conditions. Moreover, a novel fixed-time stability criterion was proposed for high-order MASs in Zhang et al. (2022), and it was guaranteed that the consensus tracking errors were driven to a small neighborhood of zero in fixed time. Although there are many works on fixed-time control, the adaptive fixed-time control of nonlinear MASs subject to actuator failures has not been fully studied, and the "explosion of complexity" problem has not been well solved.

Motivated by the above results, an adaptive fixed-time control scheme is proposed for a class of nonlinear MASs with actuator failures based on a first-order sliding mode differentiator. The main contributions of this chapter are summarized as follows.

1. The adaptive fixed-time stability criterion is extended to a more general class of nonlinear MASs with actuator faults. Compared with Liu et al. 2015, 2016; Chen et al. 2018, this chapter proposes a more helpful tool for solving the problem of approximation-based adaptive fixed-time control for nonlinear uncertain systems.
2. A DSC method employing the sliding mode differentiator is used in this chapter, it can avoid the tedious analytic computation and "explosion of complexity" problem simultaneously in the conventional backstepping method. Different from the filtering methods in Wang et al. (2022), Xia et al. (2021), a filtering error compensation mechanism is proposed in this chapter. Compared with the above compensation mechanisms, it not only makes the filtering errors bounded in fixed time but also easier to design.
3. In addition, an actuator failure model with gain fault, bias fault, and unknown control direction is constructed, and the Nussbaum function is used to build the actuator fault compensation mechanism. The proposed control strategy ensures that all error signals in MASs are bounded and the consensus is achieved in fixed time.

6.2 Preliminaries

In this section, some basic knowledge about graph theory, radial basis function (RBF) NNs, and fixed-time convergence are introduced. Some lemmas are provided for designing the following fixed-time control.

6.2.1 *Graph Theory*

Let $G = (N, V, A)$ be a digraph, where $N = \{n_1, \ldots, n_M\}$ denotes the node set, $V \subseteq N \times N$ is the edge set, and $A = [a_{hm}] \in R^{M \times M}$ is the adjacency matrix with a_{hm} being the element of A. $a_{hm} = 1$ $(h \neq m)$ if and only if the information passes from node m to h, otherwise, $a_{hm} = 0$. If there exists an edge from agent h to agent m, then $(n_h, n_m) \in V$ and $M_h = \{m \,|(n_h, n_m) \in V\}$ denotes the neighbor set of agent h. Define the in-degree matrix as $B = \text{diag}(b_1, \ldots, b_M)$ with $b_h = \sum_{m \in M_h} a_{hm}$. The Laplacian matrix L is defined by $L = B - A$. Furthermore, the digraph has a direct path between nodes n_h and n_m, if there is a sequence of edge $\{(n_h, n_r), (n_r, n_v), \ldots, (n_t, n_m)\}$. Let the leader be considered so the digraph is rewritten by the augmented graph $\bar{G} = (\bar{N}, \bar{V})$, where $\bar{N} = [0, n_1, \ldots, n_M]$ and 0 means the node of the leader. Let $c_h = 1$ if the message passes from node 0 to h, otherwise, $c_h = 0$.

6.2.2 RBFNNs

Based on Chen et al. (2018), Li and Tong (2017), RBFNNs are utilized to approximate $\phi_{h,s}\left(\bar{x}_{h,s}\right)$ over a compact set $\Omega_X \subset R^q$ as

$$\phi_{h,s}\left(\bar{x}_{h,s}\right) = W^{*T}\varphi(X) + \varepsilon(X), \quad \forall X \in \Omega_X$$

where W^* is the ideal constant weight vector which is given as

$$W^* := \arg\min_{W \in \bar{R}^l} \left\{ \sup_{X \in \Omega_X} \left|\phi_{h,s}\left(\bar{x}_{h,s}\right) - W^T\varphi(X)\right| \right\}$$

and $\varepsilon(X)$ is the approximation error. There exists a positive constant $\varepsilon^* \geq |\varepsilon(X)|$ for all $X \in \Omega_X$. The weight vector is $W = [W_1, \ldots, W_\ell]^T \in R^\ell$, the RBFNNs node number is ℓ and $\ell > 1$, the input vector is $X \in \Omega_X \subset R^q$, and the Gaussian basis function vector is $\varphi(X) = [\varphi_1(X), \ldots, \varphi_\ell(X)]^T$. Gaussian basis function is defined as

$$\varphi_i(X) = \exp\left[\frac{-(X-\mu_i)^T(X-\mu_i)}{\eta_i^2}\right], \quad i = 1, 2, \ldots, \ell$$

where $\mu_i = \left[\mu_{i1}, \ldots, \mu_{iq}\right]^T$ and η_i express the center and the width of the Gaussian function, respectively.

6.2.3 Fixed-Time Convergence

In fixed-time consensus, if there exists a bounded constant T satisfying $\lim_{t\to T} x(t) = 0$ for all time $t \geq T$, and T is not associated with initial conditions, the multi-agent systems (6.1) are fixed-time stable.

Lemma 6.1 (Zhang et al. 2022) *A system model is considered as follows*

$$\dot{x} = \phi(x), \quad x\left(t_0\right) = x_0$$

where $\phi(x) : R^n \to R^n$ denotes a smooth vector field, where x is the state vector. Define a positive definite scalar smooth function $V(x)$. If the following inequality holds,

$$\dot{V}(x) \leq -\left(c_1 V^\mu(x) + c_2 V^\beta(x)\right)^\kappa + \rho$$

where $c_1, c_2, \mu, \beta, \rho$ *and* κ *are positive constants and satisfy* $\mu\kappa \geq 1$, $0 < \beta\kappa < 1$. *Let* $\Omega_1 = \{x | V(x) \leq 1\}$, $\Omega_{a_0} = \left\{x | V(x) \leq \left(\frac{\rho}{a_0}\right)^{\frac{1}{\mu\kappa}}\right\}$, *and* $\Omega_{b_0} = \left\{x | V(x) \leq \left(\frac{\rho}{b_0}\right)^{\frac{1}{\beta\kappa}}\right\}$. *The following conclusions can be obtained.*

(1) $x(t) \in \Omega_{a_0}$ *if* Ω_{a_0} *and* Ω_{b_0} *belong to* Ω_1, *the settling time is*

$$T_s = \frac{1}{(\mu\kappa - 1)a_0} + \frac{1}{(1 - \beta\kappa)b_0}.$$

(2) $x(t) \in (\Omega_{a_0} \bigcap \Omega_{b_0})$ *if at least one of* Ω_{a_0} *and* Ω_{b_0} *is outside* Ω_1, *and the settling time is*

$$T_s = \frac{1}{(\mu\kappa - 1)a_0 \left(\frac{\rho}{b_0}\right)^{\frac{\mu\kappa - 1}{\beta\kappa}}} + \frac{\left(\frac{\rho}{b_0}\right)^{\frac{\mu\kappa - 1}{\beta\kappa}}}{(1 - \beta\kappa)b_0}$$

where

$$a_0 = \begin{cases} c_1^{\kappa}, & \kappa \geq 1 \\ 2^{\kappa-1}c_1^{\kappa}, & \kappa < 1, \end{cases} \quad b_0 = \begin{cases} c_2^{\kappa}, & \kappa \geq 1 \\ 2^{\kappa-1}c_2^{\kappa}, & \kappa < 1. \end{cases}$$

Lemma 6.2 (Zhang et al. 2022) *Let* $a_l \geq 0$, $F \in Z_+$. *For* $0 < \kappa \leq 1$, *the following inequality holds*

$$\left(\sum_{l=1}^{F} a_l\right)^{\kappa} \leq \sum_{l=1}^{F} a_l^{\kappa} \leq F^{1-\kappa}\left(\sum_{l=1}^{F} a_l\right)^{\kappa}.$$

If $\kappa > 1$, *one has*

$$F^{1-\kappa}\left(\sum_{l=1}^{F} a_l\right)^{\kappa} \leq \sum_{l=1}^{F} a_l^{\kappa}.$$

Lemma 6.3 (Wang et al. (2016)) *Define* $\tilde{\theta} = \theta - \hat{\theta}$, *where* θ *and* $\hat{\theta}$ *are two scalar functions. Let* q *and* p *be odd integers. If* $0 < c = q/p \leq 1$, *one has*

$$\tilde{\theta}\hat{\theta}^{c} \leq c_1\theta^{1+c} - c_2\tilde{\theta}^{1+c}$$

where $c_1 = \frac{1}{1+c}\left[1 - 2^{c-1} + \frac{c}{1+c} + \frac{2^{c}(1-c^2)}{1+c}\right]$, $c_2 = \frac{2^{c-1}}{1+c}\left[1 - 2^{c(c-1)}\right]$.

Lemma 6.4 (Zhang et al. 2022) *Let* $\theta > 0$ *be a constant and* $\tilde{\theta} < \theta$. *For any odd number* $p > 1$, *the following holds*

$$\tilde{\theta}(\theta - \tilde{\theta})^p \leq \theta^{p+1} - \tilde{\theta}^{p+1}.$$

Lemma 6.5 (Zhang et al. 2022) *Consider a differential equation as*

$$\dot{\hat{\theta}}(t) = -\omega\hat{\theta}^{2\mu-1}(t) - \sigma\hat{\theta}^{2\upsilon-1}(t) + \varrho(t)$$

where $\hat{\theta}(t) \in R$, ω *and* σ *are positive constants,* $\mu = p_1/q_1$, $\upsilon = q_2/p_2 > 0.5$ *with* $p_1 > q_1$ *and* $p_2 > q_2$ *being positive odd numbers,* $\varrho(t)$ *is a positive function. If* $\hat{\theta}\,(t_0) \geq 0$, $\hat{\theta}(t) \geq 0$ *holds for* $\forall t \geq t_0$.

Lemma 6.6 *Define* $\tilde{\theta} = \theta - \hat{\theta}$, *where* θ *and* $\hat{\theta}$ *are two smooth functions. Let* q *and* p *be odd integers and* $c = q/p$. *There exists a positive constant* κ *such that*

$$\kappa\tilde{\theta}(\theta - \tilde{\theta})^c \leq \theta^{1+c} - \kappa\tilde{\theta}^{1+c}.$$

Proof If $\theta = 0, \kappa\tilde{\theta}(-\tilde{\theta})^c + \kappa\tilde{\theta}^{1+c} = -\kappa\tilde{\theta}^{1+c} + \kappa\tilde{\theta}^{1+c} = 0$ with $c = q/p$, and q and p as odd integers, one has $\theta^{1+c} \geq \kappa(\tilde{\theta}(\theta - \tilde{\theta})^c + \tilde{\theta}^{1+c})$. If $\theta \neq 0$, $\theta^{1+c} > 0$ holds, and the inequality $\theta^{1+c} \geq \kappa(\tilde{\theta}(\theta - \tilde{\theta})^c + \tilde{\theta}^{1+c})$ is satisfied while κ is small enough. Therefore, $\kappa\tilde{\theta}(\theta - \tilde{\theta})^c \leq \theta^{1+c} - \kappa\tilde{\theta}^{1+c}$ always holds while κ is small enough. □

6.2.4 Nussbaum Function

The Nussbaum technique is utilized to solve the problem of unknown control coefficients (Zhang et al. 2022). The Nussbaum-type function $N(\varphi)$ is defined as

$$\lim_{s\to\infty} \sup \frac{1}{s} \int_0^s N(\varphi)d\varphi = \infty$$

$$\lim_{s\to\infty} \inf \frac{1}{s} \int_0^s N(\varphi)d\varphi = -\infty.$$

In this chapter, we choose the Nussbaum gain function as $\varphi^2 \sin(\varphi)$.

Lemma 6.7 (Li and Tong 2017) *Define* $V(t, x)$ *and* $\varphi(t)$ *as smooth functions on* $[0, t_f)$ *with* $V(t, x) \geq 0$, *and* $N(\varphi)$ *as a Nussbaum-type function. If the inequality satisfies*

$$V(t, x) \leq -EV(t, x) + (g_h N'(\varphi) + 1)\dot{\varphi} + F, \quad \forall t \in \left[0, t_f\right)$$

where E *and* F *represent positive constants.* g_h *stands for a unknown constant, then* $V(t, x)$, $\varphi(t)$ *and* $(g_h N'(\varphi) + 1)\dot{\varphi}$ *are bounded on* $\left[0, t_f\right)$.

6.3 Problem Formulation and Main Results

6.3.1 Problem Formulation

A leader-follower MAS is considered in this chapter, where one of them is the leader (marked with d) and the others are followers (marked with h from 1 to M $(M \geq 2)$). The dynamics of the followers are described as

$$\begin{aligned} \dot{x}_{h,s} &= x_{h,s+1} + \phi_{h,s}\left(\bar{x}_{h,s}\right) \\ \dot{x}_{h,n} &= \beta_h u_{hf} + \phi_{h,n}\left(x_h\right) \\ y_h &= x_{h,1}, \quad h = 1, \ldots, M \end{aligned} \tag{6.1}$$

where $x_h = [x_{h,1}, x_{h,2}, \ldots, x_{h,n}]^T \in R^n$, $\bar{x}_{h,s} = [x_{h,1}, \ldots, x_{h,s}]^T \in R^s$, $1 \leq s \leq n-1$, denote the states, $u_{hf} \in R$ and $y_h \in R$ represent the control output signal and the system output signal, respectively. β_h is an unknown constant. $\phi_{h,s}$ $(1 \leq s \leq n)$ are unknown system continuous functions and satisfy $\phi_{h,s}(0) = 0$. According to Li and Tong (2017), the actuator fault is modeled as follows:

$$u_{hf}(t) = s_h u_h(t) + w_h(t) \tag{6.2}$$

where $u_h(t)$ is a control input signal, $s_h \in (0, 1]$ is the unknown actuation effectiveness and $w_h(t)$ is a time-varying and bounded function.

Control Objective: The control objective of this chapter is to design an adaptive neural control scheme with an error compensation mechanism for nonlinear MASs subject to actuator failures and to guarantee that all signals in the closed-loop system are bounded and the consensus of MASs is achieved in fixed time.

6.3.2 Adaptive Controller Design

In this chapter, at the sth $(1 \leq s \leq n)$ step of backstepping design, RBFNNs $\Psi_{h,s}^T \Xi_{h,s}$ are utilized to approximate unknown functions. An unknown constant $\theta_{h,s} = \max\{\epsilon_{h,s}, \|\Psi_{h,s}\|\}$ is approximated by a designed adaptive parameter $\hat{\theta}_{h,s}$, and its update law $\dot{\hat{\theta}}_{h,s}$ will be given later. Symbol $\|\cdot\|$ denotes the Euclidean norm. Let the estimation error be $\tilde{\theta}_{h,s} = \theta_{h,s} - \hat{\theta}_{h,s}$. In addition, we set

$$\xi_{h,1} = \sum_{m \in M_h} a_{hm}\left(y_h - y_m\right) + c_h\left(y_h - y_d\right) \tag{6.3}$$

as the consensus error of hth follower. The output signal of the leader is defined as $y_d \in R$. Define $\xi_{h,i} = x_{h,i} - x_{h,ic}$ for $i = 2, \ldots, n$.

Step 1. For the hth follower, the Lyapunov function is designed as below

$$V_{h,1} = \frac{\xi_{h,1}^2}{2} + \frac{\tilde{\theta}_{h,1}^2}{2\eta_{h,1}} \tag{6.4}$$

with $\eta_{h,1}$ being a designed positive constant. Differentiating both sides of (6.4), we get

$$\dot{V}_{h,1} = \xi_{h,1}\dot{\xi}_{h,1} - \frac{\tilde{\theta}_{h,1}}{\eta_{h,1}}\dot{\hat{\theta}}_{h,1}. \tag{6.5}$$

From (6.1) and (6.3), one obtains

$$\begin{aligned}\dot{\xi}_{h,1} = &\ (b_h + c_h) x_{h,2} + (b_h + c_h)\phi_{h,1}\left(x_{h,1}\right) \\ &- \sum_{m\in M_h} a_{hm}\left[x_{m,2} + \phi_{m,1}\left(x_{m,1}\right)\right] - c_h\dot{y}_d \\ = &\ d_h x_{h,2} - \sum_{m\in M_h} a_{hm}x_{m,2} - c_h\dot{y}_d + \bar{\phi}_{h,1}(X_{h,1})\end{aligned} \tag{6.6}$$

where $\bar{\phi}_{h,1}(X_{h,1}) = d_h\phi_{h,1}(x_{h,1}) - \sum_{m\in M_h} a_{hm}\phi_{m,1}(x_{m,1})$ with $d_h = b_h + c_h$ and $X_{h,1} = \left[x_{h,1}, x_{m,1}\right]^T$. From the definition of graph theory, $b_h > 0$ and $c_h \in \{0, 1\}$ always hold such that $d_h > 0$. Combining (6.5) with (6.6), we get

$$\begin{aligned}\dot{V}_{h,1} = &\ \xi_{h,1}\Big(d_h x_{h,2} - \sum_{m\in M_h} a_{hm}x_{m,2} - c_h\dot{y}_d \\ &+ \bar{\phi}_{h,1}\left(X_{h,1}\right)\Big) - \frac{\tilde{\theta}_{h,1}}{\eta_{h,1}}\dot{\hat{\theta}}_{h,1}.\end{aligned} \tag{6.7}$$

In this part, RBFNNs are utilized to approximate unknown function $\bar{\phi}_{h,1}$ as

$$\bar{\phi}_{h,1}(X_{h,1}) = \Psi_{h,1}^T\Xi_{h,1}(X_{h,1}) + \delta_{h,1}(X_{h,1}) \tag{6.8}$$

where $\delta_{h,1}(X_{h,1})$ stands for the RBFNNs approximation error, and the inequality $|\delta_{h,1}(X_{h,1})| \le \epsilon_{h,1}$ is satisfied for any given accuracy $\epsilon_{h,1} > 0$. For $\forall\varsigma > 0$, one has

$$\xi_{h,1}\bar{\phi}_{h,1}\left(X_{h,1}\right) \le \xi_{h,1}\theta_{h,1}\vartheta_{h,1}\tanh\left(\frac{\xi_{h,1}\vartheta_{h,1}}{\varsigma}\right) + \varepsilon\varsigma\theta_{h,1} \tag{6.9}$$

with $\theta_{h,1} = \max\{\epsilon_{h,1}, \|\Psi_{h,1}\|\}$, $\vartheta_{h,1} = 1 + \|\Xi_{h,1}\|$, $\varepsilon = 0.2785$. Substituting (6.9) into (6.7) yields

$$\dot{V}_{h,1} \le \xi_{h,1}\Big[d_h(\xi_{h,2} + x_{h,2c}) - \sum_{m\in M_h} a_{hm}x_{m,2} - c_h\dot{y}_d$$

$$+\theta_{h,1}\vartheta_{h,1}\tanh\left(\frac{\xi_{h,1}\vartheta_{h,1}}{\varsigma}\right)\Bigg]-\frac{\tilde{\theta}_{h,1}}{\eta_{h,1}}\dot{\hat{\theta}}_{h,1}+\varepsilon\varsigma\theta_{h,1}. \tag{6.10}$$

Therefore, the control signal $x_{h,2c}$ of the hth subsystem is

$$x_{h,2c}=\frac{1}{d_h}\Bigg[-k_{h,11}\xi_{h,1}^{2\mu-1}+\sum_{m\in M_h}a_{hm}x_{m,2}+c_h\dot{y}_d$$
$$+\alpha_{h,1}-\hat{\theta}_{h,1}\vartheta_{h,1}\tanh\left(\frac{\xi_{h,1}\vartheta_{h,1}}{\varsigma}\right)\Bigg] \tag{6.11}$$

where $\mu=p_1/q_1>1$, p_1,q_1 are odd numbers and $p_1>q_1$, $\alpha_{h,1}$ is designed as below to ensure that $\dot{x}_{h,2c}$ exists in the next step.

$$\alpha_{h,1}=\begin{cases} f\left(\xi_{h,1}\right), & \left\|\xi_{h,1}\right\|<\epsilon_{h,1}\\ -k_{h,12}\xi_{h,1}^{2\beta-1}, & \left\|\xi_{h,1}\right\|\geq\epsilon_{h,1}\end{cases} \tag{6.12}$$

where $f\left(\xi_{h,1}\right)=g_{h,11}\xi_{h,1}+g_{h,12}\xi_{h,1}^3$, $k_{h,11}$ and $k_{h,12}$ are positive design parameters. Let $\beta=q_2/p_2$ $(0.5<\beta<1)$, where p_2 and q_2 are odd numbers and satisfy $p_2>q_2$. Similar to Zhang et al. (2022), the curve fitting method is utilized to ensure that $\alpha_{h,1}$ is derivable when $\xi_{h,1}=\pm\epsilon_{h,1}$, such that the following equation holds

$$f(\xi_{h,1})=-k_{h,12}\xi_{h,1}^{2\nu-1}\mid_{\xi_{h,1}=\pm\epsilon_{h,1}}$$
$$\frac{\partial f}{\partial\xi_{h,1}}=-k_{h,12}(2\nu-1)\xi_{h,1}^{2\nu-2}\mid_{\xi_{h,1}=\pm\epsilon_{h,1}}$$

where $g_{h,11}=-k_{h,12}\epsilon_{h,1}^{2\beta-2}(2-\beta)$ and $g_{h,12}=-k_{h,12}\epsilon_{h,1}^{2\beta-4}(\beta-1)$. Substituting $x_{h,2c}$ into $\dot{V}_{h,1}$, we get

$$\dot{V}_{h,1}\leq-k_{h,11}\xi_{h,1}^{2\mu}+\tilde{\theta}_{h,1}\xi_{h,1}\vartheta_{h,1}\tanh\left(\frac{\xi_{h,1}\vartheta_{h,1}}{\varsigma}\right)$$
$$+\xi_{h,1}\alpha_{h,1}-\frac{\tilde{\theta}_{h,1}}{\eta_{h,1}}\dot{\hat{\theta}}_{h,1}+\varepsilon\varsigma\theta_{h,1}+d_h\xi_{h,1}\xi_{h,2}. \tag{6.13}$$

For $\left\|\xi_{h,1}\right\|\geq\epsilon_{h,1}$, in light of (6.12), we have

$$\xi_{h,1}\alpha_{h,1}=-k_{h,12}\xi_{h,1}^{2\beta}. \tag{6.14}$$

For $\left\|\xi_{h,1}\right\|<\epsilon_{h,1}$, under the condition of $\xi_{h,1}<\epsilon_{h,1}<1$, we have

$$\xi_{h,1}\alpha_{h,1}=-k_{h,12}\epsilon_{h,1}^{2\beta-2}\left((2-\beta)\xi_{h,1}^2+(\beta-1)\epsilon_{h,1}^{-2}\xi_{h,1}^4\right)$$
$$\leq-k_{h,12}\epsilon_{h,1}^{2\beta-2}\xi_{h,1}^2-k_{h,12}(1-\beta)\epsilon_{h,1}^{2\beta-2}\xi_{h,1}^2$$

$$
\begin{aligned}
&+k_{h,12}(1-\beta)\epsilon_{h,1}^{2\beta-4}\xi_{h,1}^{4} \\
\leq &-k_{h,12}\xi_{h,1}^{2\beta}.
\end{aligned} \tag{6.15}
$$

From (6.14) and (6.15), $\xi_{h,1}\alpha_{h,1} \leq -k_{h,12}\xi_{h,1}^{2\beta}$ is satisfied. Therefore, (6.13) can be rewritten as below

$$
\begin{aligned}
\dot{V}_{h,1} \leq &-k_{h,11}\xi_{h,1}^{2\mu}-k_{h,12}\xi_{h,1}^{2\beta}+\varepsilon\varsigma\theta_{h,1}+d_h\xi_{h,1}\xi_{h,2} \\
&-\frac{\tilde{\theta}_{h,1}}{\eta_{h,1}}\left(\dot{\hat{\theta}}_{h,1}-\eta_{h,1}\xi_{h,1}\vartheta_{h,1}\tanh\left(\frac{\xi_{h,1}\vartheta_{h,1}}{\varsigma}\right)\right).
\end{aligned} \tag{6.16}
$$

According to (6.16), the adaptive update law $\dot{\hat{\theta}}_{h,1}$ is given by

$$
\begin{aligned}
\dot{\hat{\theta}}_{h,1}= &-\sigma_{h,11}\hat{\theta}_{h,1}^{2\mu-1}-\sigma_{h,12}\hat{\theta}_{h,1}^{2\beta-1} \\
&+\eta_{h,1}\xi_{h,1}\vartheta_{h,1}\tanh\left(\frac{\xi_{h,1}\vartheta_{h,1}}{\varsigma}\right)
\end{aligned} \tag{6.17}
$$

where $\sigma_{h,11}$ and $\sigma_{h,12}$ are positive design parameters. Thus, substituting (6.17) into (6.16) has

$$
\begin{aligned}
\dot{V}_{h,1} \leq &-k_{h,11}\xi_{h,1}^{2\mu}-k_{h,12}\xi_{h,1}^{2\beta}+\frac{\sigma_{h,11}}{\eta_{h,1}}\tilde{\theta}_{h,1}\hat{\theta}_{h,1}^{2\mu-1} \\
&+\frac{\sigma_{h,12}}{\eta_{h,1}}\tilde{\theta}_{h,1}\hat{\theta}_{h,1}^{2\beta-1}+\varepsilon\varsigma\theta_{h,1}+d_h\xi_{h,1}\xi_{h,2}.
\end{aligned} \tag{6.18}
$$

Step s ($2 \leq s \leq n-1$). At this step, the Lyapunov function $V_{h,s}$ is constructed as

$$
V_{h,s}=V_{h,s-1}+\frac{\xi_{h,s}^2}{2}+\frac{\tilde{\theta}_{h,s}^2}{2\eta_{h,s}} \tag{6.19}
$$

where $\eta_{h,s}$ is a positive design parameter. Then, differentiating both sides of (6.19), we get

$$
\begin{aligned}
\dot{V}_{h,s}= &\ \dot{V}_{h,s-1}+\xi_{h,s}(\xi_{h,s+1}+x_{h,s+1c}+\phi_{h,s}(X_{h,s})-\dot{x}_{h,sc} \\
&+\ell_{h,s}\xi_{h,s-1})-\frac{\tilde{\theta}_{h,s}}{\eta_{h,s}}\dot{\hat{\theta}}_{h,s}-\ell_{h,s}\xi_{h,s-1}\xi_{h,s}
\end{aligned} \tag{6.20}
$$

where $\ell_{h,s}=d_h$ while $s=2$, and $\ell_{h,s}=1$ while $2<s\leq n$. Considering the problem of tedious calculations, similar to Chen and Ge (2015), Wang and Chen (2020), Lin et al. (2019), the first-order sliding mode differentiator is used to approximate $\dot{x}_{h,sc}$,

$$
\begin{aligned}
\dot{\varrho}_{h,s0} =& l_{h,s} \\
l_{h,s} =& -R_{h,s0}\left|\varrho_{h,s0}-x_{h,sc}\right|^{\frac{1}{2}}\operatorname{sign}\left(\varrho_{h,s0}-x_{h,sc}\right)+\varrho_{h,s1} \\
\dot{\varrho}_{h,s1} =& -R_{h,s1}\operatorname{sign}\left(\varrho_{h,s1}-l_{h,s}\right)
\end{aligned} \tag{6.21}
$$

where $\varrho_{h,sj}$ and $l_{h,s}$ are system states of the differentiator, and $R_{h,sj}$ are positive design constants ($j = 0, 1$). Meanwhile, $\varrho_{h,s0}$ and $l_{h,s}$ are the estimations of $x_{h,sc}$ and $\dot{x}_{h,sc}$, respectively. Therefore, we have

$$
\dot{x}_{h,sc} = l_{h,s} + p_{h,s} \tag{6.22}
$$

where $p_{h,s}$ stands for the estimated error of differentiator.

Remark 6.1 In addition, to ensure that the estimated error $p_{h,s}$ is fixed-time bounded, a novel filtering error compensation mechanism is designed as follows:

$$
p_{h,s} = \hat{p}_{h,s} + \tilde{p}_{h,s} \tag{6.23}
$$

where $\tilde{p}_{h,s}$ is the compensation error and $\hat{p}_{h,s}$ is designed as an adaptive law to approximate filtering error $p_{h,s}$.

Therefore, (6.20) can be rewritten as

$$
\begin{aligned}
\dot{V}_{h,s} =& \dot{V}_{h,s-1} + \xi_{h,s}(\xi_{h,s+1} + x_{h,s+1c} + \phi_{h,s}(X_{h,s}) - l_{h,s} \\
& - \hat{p}_{h,s} - \tilde{p}_{h,s} + \ell_{h,s}\xi_{h,s-1}) - \frac{\tilde{\theta}_{h,s}}{\eta_{h,s}}\dot{\hat{\theta}}_{h,s} \\
& - \ell_{h,s}\xi_{h,s-1}\xi_{h,s}.
\end{aligned} \tag{6.24}
$$

Utilizing RBFNNs to estimate unknown function $\phi_{h,s}(X_{h,s})$, for $\forall\varsigma > 0$, we have

$$
\xi_{h,s}\phi_{h,s}\left(X_{h,s}\right) \leq \xi_{h,s}\theta_{h,s}\vartheta_{h,s}\tanh\left(\frac{\xi_{h,s}\vartheta_{h,s}}{\varsigma}\right) + \varepsilon\varsigma\theta_{h,s} \tag{6.25}
$$

where $\theta_{h,s} = \max\{\epsilon_{h,s}, \|\Psi_{h,s}\|\}$ and $\vartheta_{h,s} = 1 + \|\Xi_{h,s}\|$. For the error function $\delta_{h,s}(X_{h,s})$, there exists $|\delta_{h,s}(X_{h,s})| \leq \epsilon_{h,s}$, where $\epsilon_{h,s} > 0$ is a given accuracy. Substituting (6.25) into (6.24), then, define

$$
\begin{aligned}
x_{h,s+1c} =& -k_{h,s1}\xi_{h,s}^{2\mu-1} - \hat{\theta}_{h,s}\vartheta_{h,s}\tanh\left(\frac{\xi_{h,s}\vartheta_{h,s}}{\varsigma}\right) \\
& + l_{h,s} + \hat{p}_{h,s} - \ell_{h,s}\xi_{h,s-1} + \alpha_{h,s}
\end{aligned} \tag{6.26}
$$

$$
\alpha_{h,s} = \begin{cases} f\left(\xi_{h,s}\right), & \left\|\xi_{h,s}\right\| < \epsilon_{h,s} \\ -k_{h,s2}\xi_{h,s}^{2\beta-1}, & \left\|\xi_{h,s}\right\| \geq \epsilon_{h,s} \end{cases} \tag{6.27}
$$

where $f\left(\xi_{h,s}\right) = g_{h,s1}\xi_{h,s} + g_{h,s2}\xi_{h,s}^3$, $k_{h,s1}$ and $k_{h,s2}$ are positive design parameters. Let $g_{h,s1} = -k_{h,s2}\epsilon_{h,s}^{2\beta-2}(2-\beta)$ and $g_{h,s2} = -k_{h,s2}\epsilon_{h,s}^{2\beta-4}(\beta-1)$, such that $\xi_{h,s}\alpha_{h,s} \leq -k_{h,s2}\xi_{h,s}^{2\beta}$ holds. Substituting (6.25) and (6.26) into (6.24), we have

$$\begin{aligned}\dot{V}_{h,s} \leq\ & \dot{V}_{h,s-1} - k_{h,s1}\xi_{h,s}^{2\mu} - k_{h,s2}\xi_{h,s}^{2\beta} + \xi_{h,s}\xi_{h,s+1} \\ & - \frac{\tilde{\theta}_{h,s}}{\eta_{h,s}}\left(\dot{\hat{\theta}}_{h,s} - \eta_{h,s}\xi_{h,s}\vartheta_{h,s}\tanh\left(\frac{\xi_{h,s}\vartheta_{h,s}}{\varsigma}\right)\right) \\ & - \ell_{h,s}\xi_{h,s-1}\xi_{h,s} - \tilde{p}_{h,s}\xi_{h,s} + \varepsilon\varsigma\theta_{h,s}. \end{aligned} \tag{6.28}$$

According to (6.28), the adaptive update law $\dot{\hat{\theta}}_{h,s}$ is given by

$$\dot{\hat{\theta}}_{h,s} = -\sigma_{h,s1}\hat{\theta}_{h,s}^{2\mu-1} - \sigma_{h,s2}\hat{\theta}_{h,s}^{2\beta-1} + \eta_{h,s}\xi_{h,s}\vartheta_{h,s}\tanh\left(\frac{\xi_{h,s}\vartheta_{h,s}}{\varsigma}\right) \tag{6.29}$$

where $\sigma_{h,s1}$ and $\sigma_{h,s2}$ are positive design parameters. Therefore, the formula (6.28) is rewritten as

$$\begin{aligned}\dot{V}_{h,s} \leq & -\sum_{m=1}^{s}\left(k_{h,m1}\xi_{h,m}^{2\mu} + k_{h,m2}\xi_{h,m}^{2\beta}\right) \\ & + \sum_{m=1}^{s}\frac{\tilde{\theta}_{h,m}}{\eta_{h,m}}\left(\sigma_{h,m1}\hat{\theta}_{h,m}^{2\mu-1} + \sigma_{h,m2}\hat{\theta}_{h,m}^{2\beta-1}\right) \\ & + \sum_{m=1}^{s}\varepsilon\varsigma\theta_{h,m} + \xi_{h,s}\xi_{h,s+1} - \sum_{m=2}^{s}\tilde{p}_{h,m}\xi_{h,m}. \end{aligned} \tag{6.30}$$

Step n. For a positive constant $\eta_{h,n}$, the Lyapunov function $V_{h,n}$ is designed as

$$V_{h,n} = V_{h,n-1} + \frac{\xi_{h,n}^2}{2} + \frac{\tilde{\theta}_{h,n}^2}{2\eta_{h,n}}. \tag{6.31}$$

Utilizing RBFNN to approximate $\phi_{h,n}(X_{h,n})$, $\dot{V}_{h,n}$ is shown as follows:

$$\begin{aligned}\dot{V}_{h,n} \leq\ & \dot{V}_{h,n-1} + \xi_{h,n}\big(g_h u_h + \theta_{h,n}\vartheta_{h,n}\tanh\left(\frac{\xi_{h,n}\vartheta_{h,n}}{\varsigma}\right) \\ & - l_{h,n} - \hat{p}_{h,n} + \ell_{h,n}\xi_{h,n-1} + \frac{\xi_{h,n}}{2}\big) - \frac{\tilde{\theta}_{h,n}}{\eta_{h,n}}\dot{\hat{\theta}}_{h,n} \\ & - \ell_{h,n}\xi_{h,n-1}\xi_{h,n} + \varepsilon\varsigma\theta_{h,n} + \frac{\bar{w}_h^2}{2} - \tilde{p}_{h,n}\xi_{h,n} \end{aligned} \tag{6.32}$$

where $g_h = \beta_h s_h$, $\theta_{h,n} = \max\{\epsilon_{h,n}, \|\Psi_{h,n}\|\}$ and $\vartheta_{h,n} = 1 + \|\Xi_{h,n}\|$, and $\bar{w}_h \geq |s_h w_h|$ $(1 \leq h \leq M)$. According to (6.32), the control input signal u_h and adaptive update law $\dot{\hat{\theta}}_{h,n}$ are given as

$$\dot{\varphi}_h = \xi_{h,n}\Big(k_{h,n1}\xi_{h,n}^{2\mu-1} + k_{h,n2}\xi_{h,n}^{2\beta-1} - l_{h,n} - \hat{p}_{h,n} + \hat{\theta}_{h,n}\vartheta_{h,n}\tanh\left(\frac{\xi_{h,n}\vartheta_{h,n}}{\varsigma}\right) + \ell_{h,n}\xi_{h,n-1} + \frac{\xi_{h,n}}{2}\Big) \tag{6.33}$$

$$u_h = N(\varphi_h)\Big(k_{h,n1}\xi_{h,n}^{2\mu-1} + k_{h,n2}\xi_{h,n}^{2\beta-1} - l_{h,n} - \hat{p}_{h,n} + \hat{\theta}_{h,n}\vartheta_{h,n}\tanh\left(\frac{\xi_{h,n}\vartheta_{h,n}}{\varsigma}\right) + \ell_{h,n}\xi_{h,n-1} + \frac{\xi_{h,n}}{2}\Big) \tag{6.34}$$

$$\dot{\hat{\theta}}_{h,n} = -\sigma_{h,n1}\hat{\theta}_{h,n}^{2\mu-1} - \sigma_{h,n2}\hat{\theta}_{h,n}^{2\beta-1} + \eta_{h,n}\xi_{h,n}\vartheta_{h,n}\tanh\left(\frac{\xi_{h,n}\vartheta_{h,n}}{\varsigma}\right) \tag{6.35}$$

where $k_{h,nj}$ and $\sigma_{h,nj}$ are positive design constants ($j = 1, 2$), and the Nussbaum function $N(\varphi_h)$ is chosen as $\varphi_h^2 \sin(\varphi_h)$. Substituting (6.33)–(6.35) into (6.32), we have

$$\dot{V}_{h,n} \le -\sum_{s=1}^{n}\left(k_{h,s1}\xi_{h,s}^{2\mu} + k_{h,s2}\xi_{h,s}^{2\beta}\right) + \sum_{s=1}^{n}\frac{\tilde{\theta}_{h,s}}{\eta_{h,s}}\left(\sigma_{h,s1}\hat{\theta}_{h,s}^{2\mu-1} + \sigma_{h,s2}\hat{\theta}_{h,s}^{2\beta-1}\right) - \sum_{s=2}^{n}\tilde{p}_{h,s}\xi_{h,s} + \left(g_h N(\varphi_h) + 1\right)\dot{\varphi}_h + \sum_{m=1}^{n}\varepsilon\varsigma\theta_{h,m} + \frac{\bar{w}_h^2}{2}. \tag{6.36}$$

Define $D = \left(g_h N(\varphi_h) + 1\right)\dot{\varphi}_h$, such that (6.36) can be rewritten as:

$$\dot{V}_{h,n} \le -\sum_{s=1}^{n}\left(k_{h,s1}\xi_{h,s}^{2\mu} + k_{h,s2}\xi_{h,s}^{2\beta}\right) + \sum_{s=1}^{n}\frac{\tilde{\theta}_{h,s}}{\eta_{h,s}}\left(\sigma_{h,s1}\hat{\theta}_{h,s}^{2\mu-1} + \sigma_{h,s2}\hat{\theta}_{h,s}^{2\beta-1}\right) - \sum_{s=2}^{n}\tilde{p}_{h,s}\xi_{h,s} + \sum_{m=1}^{n}\varepsilon\varsigma\theta_{h,m} + \frac{\bar{w}_h^2}{2} + D. \tag{6.37}$$

Based on Lemma 6.3, we have

$$\tilde{\theta}_{h,s}\hat{\theta}_{h,s}^{2\beta-1} \le c_1\theta_{h,s}^{2\beta} - c_2\tilde{\theta}_{h,s}^{2\beta} \tag{6.38}$$

where $c_1 = \frac{1}{2\beta}\left(1 - 2^{2\beta-2} + \frac{2\beta-1}{2\beta} + \frac{2^{2\beta-1}(1-(2\beta-1)^2)}{2\beta}\right)$, $c_2 = \frac{2^{2\beta-2}}{2\beta}\left(1 - 2^{(2\beta-1)(2\beta-2)}\right)$. Before utilizing Lemma 6.4, $\hat{\theta}_{h,s} = \theta_{h,s} - \tilde{\theta}_{h,s} > 0$ should be guaranteed. For $\hat{\theta}_{h,s}(0) \ge 0$ and $\forall t > 0$, Lemma 6.5 shows that $\hat{\theta}_{h,s}(t) > 0$ holds with adaptive law $\dot{\hat{\theta}}_{h,s}$ (6.17), (6.29) and (6.35). Therefore, from Lemma 6.4, we can get

$$\tilde{\theta}_{h,s}\hat{\theta}_{h,s}^{2\mu-1} = \tilde{\theta}_{h,s}\left(\theta_{h,s} - \tilde{\theta}_{h,s}\right)^{2\mu-1} \le \theta_{h,s}^{2\mu} - \tilde{\theta}_{h,s}^{2\mu}. \tag{6.39}$$

Substituting (6.38) and (6.39) into (6.37) yields

$$
\begin{aligned}
\dot{V}_{h,n} \leq & -\sum_{s=1}^{n} 2^{\mu} k_{h,s1} \left(\frac{\xi_{h,s}^2}{2} \right)^{\mu} - \sum_{s=1}^{n} 2^{\beta} k_{h,s2} \left(\frac{\xi_{h,s}^2}{2} \right)^{\beta} \\
& - \sum_{s=1}^{n} 2^{\mu} \eta_{h,s}^{\mu-1} \sigma_{h,s1} \left(\frac{\tilde{\theta}_{h,s}^2}{2\eta_{h,s}} \right)^{\mu} - \sum_{s=1}^{n} 2^{\beta} c_2 \eta_{h,s}^{\beta-1} \sigma_{h,s2} \left(\frac{\tilde{\theta}_{h,s}^2}{2\eta_{h,s}} \right)^{\beta} \\
& - \sum_{s=2}^{n} \tilde{p}_{h,s} \xi_{h,s} + r_h + D
\end{aligned} \tag{6.40}
$$

where $r_h = \sum_{s=1}^{n} \frac{\sigma_{h,s1}}{\eta_{h,s}} \theta_{h,s}^{2\mu} + \sum_{s=1}^{n} \frac{\sigma_{h,s2}}{\eta_{h,s}} c_1 \theta_{h,s}^{2\beta} + \sum_{m=1}^{n} \varepsilon \varsigma \theta_{h,m} + \frac{\bar{w}_h^2}{2}$.

Theorem 6.1 *For the MASs* (6.1) *with actuator faults* (6.2)*, under the first-order sliding mode differentiator* (6.21)*, the error compensation mechanism* (6.23)*, the control input signal* (6.34)*, adaptive laws* $\dot{\hat{\theta}}_{h,s}$*, and the adaptive update laws*

$$
\dot{\hat{p}}_{h,s} = -\gamma_{h,s} \xi_{h,s} - \tau_{h,s1} \hat{p}_{h,s}^{2\mu-1} - \tau_{h,s2} \hat{p}_{h,s}^{2\beta-1}, \quad s = 2, \ldots, n \tag{6.41}
$$

where $\gamma_{h,s}$ *and* $\tau_{h,sj} (j = 1, 2)$ *are positive design parameters, it can guarantee that the tracking errors* $\xi_{h,1} (h = 1, \ldots, M)$ *converge to a small neighborhood of origin and all signals are fixed-time bounded in the closed-loop system.*

Proof For the compensating mechanism (6.23), choose the Lyapunov function $\bar{V}_n$ as

$$
\bar{V}_h = \frac{1}{2} \sum_{s=2}^{n} \frac{\tilde{p}_{h,s}^2}{\gamma_{h,s}}. \tag{6.42}
$$

Differentiating $\bar{V}_h$, one gets

$$
\begin{aligned}
\dot{\bar{V}}_h &= -\sum_{s=2}^{n} \frac{\tilde{p}_{h,s}}{\gamma_{h,s}} \dot{\hat{p}}_{h,s} \\
&= -\sum_{s=2}^{n} \frac{\tilde{p}_{h,s}}{\gamma_{h,s}} (\dot{\hat{p}}_{h,s} + \gamma_{h,s} \xi_{h,s}) + \sum_{s=2}^{n} \tilde{p}_{h,s} \xi_{h,s}.
\end{aligned} \tag{6.43}
$$

Substituting (6.41) into (6.43), we have

$$
\dot{\bar{V}}_h = \sum_{s=2}^{n} \frac{\tilde{p}_{h,s}}{\gamma_{h,s}} (\tau_{h,s1} \hat{p}_{h,s}^{2\mu-1} + \tau_{h,s2} \hat{p}_{h,s}^{2\beta-1}) + \sum_{s=2}^{n} \tilde{p}_{h,s} \xi_{h,s}. \tag{6.44}
$$

Define $\kappa_{h,s1}$ and $\kappa_{h,s2}$ as positive design parameters. By utilizing Lemma 6.6, we get

$$
\kappa_{h,s1} \tilde{p}_{h,s} (p_{h,s} - \tilde{p}_{h,s})^{2\mu-1} \leq p_{h,s}^{2\mu} - \kappa_{h,s1} \tilde{p}_{h,s}^{2\mu}
$$

$$\kappa_{h,s2}\tilde{p}_{h,s}(p_{h,s}-\tilde{p}_{h,s})^{2\beta-1} \leq p_{h,s}^{2\mu} - \kappa_{h,s2}\tilde{p}_{h,s}^{2\beta}. \tag{6.45}$$

Substituting (6.45) into (6.44), $\dot{\bar{V}}_h$ is rewritten as

$$\begin{aligned}
\dot{\bar{V}}_h \leq & \sum_{s=2}^{n} \frac{\tau_{h,s1}}{\kappa_{h,s1}\gamma_{h,s}} \kappa_{h,s1}\tilde{p}_{h,s}(p_{h,s}-\tilde{p}_{h,s})^{2\mu-1} \\
& + \sum_{s=2}^{n} \frac{\tau_{h,s2}}{\kappa_{h,s2}\gamma_{h,s}} \kappa_{h,s2}\tilde{p}_{h,s}(p_{h,s}-\tilde{p}_{h,s})^{2\beta-1} + \sum_{s=2}^{n}\tilde{p}_{h,s}\xi_{h,s} \\
\leq & \sum_{s=2}^{n} \frac{\tau_{h,s1}}{\kappa_{h,s1}\gamma_{h,s}} (p_{h,s}^{2\mu} - \kappa_{h,s1}\tilde{p}_{h,s}^{2\mu}) \\
& + \sum_{s=2}^{n} \frac{\tau_{h,s2}}{\kappa_{h,s2}\gamma_{h,s}} (p_{h,s}^{2\beta} - \kappa_{h,s2}\tilde{p}_{h,s}^{2\beta}) + \sum_{s=2}^{n}\tilde{p}_{h,s}\xi_{h,s}.
\end{aligned} \tag{6.46}$$

Choose $V_h = V_{h,n} + \bar{V}_h$ as the Lyapunov function candidate of hth MASs (6.1). Differentiating V_h, $\dot{V}_h$ can be calculated as

$$\begin{aligned}
\dot{V}_h \leq & -\sum_{s=1}^{n} 2^{\mu}k_{h,s1}\left(\frac{\xi_{h,s}^2}{2}\right)^{\mu} - \sum_{s=1}^{n} 2^{\beta}k_{h,s2}\left(\frac{\xi_{h,s}^2}{2}\right)^{\beta} \\
& -\sum_{s=1}^{n}\left(2^{\mu}\eta_{h,s}^{\mu-1}\sigma_{h,s1}\left(\frac{\tilde{\theta}_{h,s}^2}{2\eta_{h,s}}\right)^{\mu} + 2^{\beta}c_2\eta_{h,s}^{\beta-1}\sigma_{h,s2}\left(\frac{\tilde{\theta}_{h,s}^2}{2\eta_{h,s}}\right)^{\beta}\right) \\
& -\sum_{s=2}^{n}\left(2^{\mu}\gamma_{h,s}^{\mu-1}\tau_{h,s1}\left(\frac{\tilde{p}_{h,s}^2}{2\gamma_{h,s}}\right)^{\mu} + 2^{\beta}\gamma_{h,s}^{\beta-1}\tau_{h,s2}\left(\frac{\tilde{p}_{h,s}^2}{2\gamma_{h,s}}\right)^{\beta}\right) + \bar{r}_h + D \\
\leq & -\varpi_{h,1}\left[\sum_{s=1}^{n}\left(\frac{\xi_{h,s}^2}{2}\right)^{\mu} + \sum_{s=1}^{n}\left(\frac{\tilde{\theta}_{h,s}^2}{2\eta_{h,s}}\right)^{\mu} + \sum_{s=2}^{n}\left(\frac{\tilde{p}_{h,s}^2}{2\gamma_{h,s}}\right)^{\mu}\right] \\
& -\varpi_{h,2}\left[\sum_{s=1}^{n}\left(\frac{\xi_{h,s}^2}{2}\right)^{\beta} + \sum_{s=1}^{n}\left(\frac{\tilde{\theta}_{h,s}^2}{2\eta_{h,s}}\right)^{\beta} + \sum_{s=2}^{n}\left(\frac{\tilde{p}_{h,s}^2}{2\gamma_{h,s}}\right)^{\beta}\right] + \bar{r}_h + D
\end{aligned} \tag{6.47}$$

where

$$\varpi_{h,1} = \min\left\{2^{\mu}k_{h,s1},\, 2^{\mu}\eta_{h,s}^{\mu-1}\sigma_{h,s1},\, 2^{\mu}\gamma_{h,s}^{\mu-1}\tau_{h,s1}\right\}$$

$$\varpi_{h,2} = \min\left\{2^{\beta}k_{h,s2},\, 2^{\beta}c_2\eta_{h,s}^{\beta-1}\sigma_{h,s2},\, 2^{\beta}\gamma_{h,s}^{\beta-1}\tau_{h,s2}\right\}$$

and $\bar{r}_h = \sum_{s=2}^{n}\frac{\tau_{h,s1}}{\kappa_{h,s1}\gamma_{h,s}}p_{h,s}^{2\mu} + \sum_{s=2}^{n}\frac{\tau_{h,s2}}{\kappa_{h,s2}\gamma_{h,s}}p_{h,s}^{2\beta} + r_h$. Applying Lemma 6.2 to (6.47), one shows

$$
\begin{aligned}
&-\varpi_{h,1}\left[\sum_{s=1}^{n}\left(\frac{\xi_{h,s}^2}{2}\right)^{\mu}+\sum_{s=1}^{n}\left(\frac{\tilde{\theta}_{h,s}^2}{2\eta_{h,s}}\right)^{\mu}+\sum_{s=2}^{n}\left(\frac{\tilde{p}_{h,s}^2}{2\gamma_{h,s}}\right)^{\mu}\right]\\
\le&-\varpi_{h,1}n^{1-\mu}\left[\left(\sum_{s=1}^{n}\frac{\xi_{h,s}^2}{2}\right)^{\mu}+\left(\sum_{s=1}^{n}\frac{\tilde{\theta}_{h,s}^2}{2\eta_{h,s}}\right)^{\mu}\right]\\
&-\varpi_{h,1}(n-1)^{1-\mu}\left(\sum_{s=2}^{n}\frac{\tilde{p}_{h,s}^2}{2\gamma_{h,s}}\right)^{\mu}\\
\le&-\varpi_{h,1}n^{1-\mu}\left[\left(\sum_{s=1}^{n}\frac{\xi_{h,s}^2}{2}\right)^{\mu}+\left(\sum_{s=1}^{n}\frac{\tilde{\theta}_{h,s}^2}{2\eta_{h,s}}\right)^{\mu}+\left(\sum_{s=2}^{n}\frac{\tilde{p}_{h,s}^2}{2\gamma_{h,s}}\right)^{\mu}\right]\\
\le&-\bar{\varpi}_{h,1}V_h^{\mu}
\end{aligned}
\tag{6.48}
$$

where $\bar{\varpi}_{h,1}=\varpi_{h,1}n^{1-\mu}3^{1-\mu}$, and

$$
\begin{aligned}
&-\varpi_{h,2}\left[\sum_{s=1}^{n}\left(\frac{\xi_{h,s}^2}{2}\right)^{\beta}+\sum_{s=1}^{n}\left(\frac{\tilde{\theta}_{h,s}^2}{2\eta_{h,s}}\right)^{\beta}+\sum_{s=2}^{n}\left(\frac{\tilde{p}_{h,s}^2}{2\gamma_{h,s}}\right)^{\beta}\right]\\
\le&-\varpi_{h,2}\left[\left(\sum_{s=1}^{n}\frac{\xi_{h,s}^2}{2}\right)^{\beta}+\left(\sum_{s=1}^{n}\frac{\tilde{\theta}_{h,s}^2}{2\eta_{h,s}}\right)^{\beta}+\left(\sum_{s=2}^{n}\frac{\tilde{p}_{h,s}^2}{2\gamma_{h,s}}\right)^{\beta}\right]\\
\le&-\varpi_{h,2}V_h^{\beta}.
\end{aligned}
\tag{6.49}
$$

Further, substituting (6.48) and (6.49) into (6.47), we have

$$
\dot{V}_h\le-\bar{\varpi}_{h,1}V_h^{\mu}-\varpi_{h,2}V_h^{\beta}+\bar{r}_h+D. \tag{6.50}
$$

From (6.50), on one hand, if $V_h\ge 1$, one has

$$
\begin{aligned}
\dot{V}_h\le&-\bar{\varpi}_{h,1}V_h^{\mu}-\varpi_{h,2}V_h^{\beta}+\bar{r}_h+D\\
<&-\bar{\varpi}_{h,1}V_h+\bar{r}_h+D.
\end{aligned}
\tag{6.51}
$$

On the other hand, if $0\le V_h<1$, we get $\dot{V}_h\le-\varpi_{h,2}V_h+\bar{r}_h+D$. By using Lemma 6.7, $|D|$ is bounded. Define $\bar{\lambda}_h$ as a positive constant and let $|D|<\bar{\lambda}_h$. For $\bar{D}=\bar{r}_h+\bar{\lambda}_h$, we get

$$
\dot{V}_h\le-\bar{\varpi}_{h,1}V_h^{\mu}-\varpi_{h,2}V_h^{\beta}+\bar{D}. \tag{6.52}
$$

Let $\kappa=1$, $\Omega_{\xi_h}=\left\{\xi_h|V_h(\xi_{h,s})\le\left(\frac{r_h}{\bar{\varpi}_{h,1}}\right)^{\frac{1}{\mu}}\right\}$, $\Omega_{\varepsilon_h}=\left\{\xi_h|\xi_{h,s}\le\varepsilon_{h,s}\right\}$ and $\Omega_h=\max\{\Omega_{\xi_h},\Omega_{\varepsilon_h}\}$. From Lemma 6.1, the inequality (6.52) shows that the tracking error $\xi_{h,1}$ and all error signals $\xi_{h,s}$ will converge to Ω_h in fixed time. □

Remark 6.2 Compared with Lemmas 6.3, 6.4 and 6.6 can be used in the both conditions $0 < c \leq 1$ and $p > 1$. Moreover, the limitations where $\theta > 0$ and $\tilde{\theta} < \theta$ are removed. However, utilizing Lemma 6.6 will create a lager $\bar{D}$ in (6.52), which makes the stability analysis of closed-loop system more conservative.

6.4 Simulation Examples

In this section, two examples are provided to verify the effectiveness of fixed-time command filtering adaptive neural controller. In simulation examples, the communication graph G of MASs is shown in Fig. 6.1. Consider the MASs with a leader and 4 followers.

Example 6.1 Consider the model of $h^{th}(h = 1, \ldots, 4)$ agent

$$\begin{aligned} \dot{x}_{h,1} &= x_{h,2} + x_{h,1}\sin(x_{h,1}^2) \\ \dot{x}_{h,2} &= u_{hf} + x_{h,1}x_{h,2}^2 \\ y_h &= x_{h,1} \end{aligned}$$

where $u_{if} = u_i$ $(i = 1, 2, 3)$, and

$$u_{4f} = \begin{cases} u_4, & t < 20 \\ -0.8u_4 + 0.2\cos(t), & t \geq 20. \end{cases}$$

The trajectory of leader is defined as $y_d = \sin(0.4t) + \sin(0.8t)$, the filter parameters are $R_{h,10} = 20$, $R_{h,11} = 5$ and other parameters are shown in Table 6.1.

Simulation results are shown in Figs. 6.2, 6.3, 6.4, 6.5, 6.6, and 6.7. The trajectories of output signal y_h and reference signal y_d are shown in Fig. 6.2. Figure 6.3 shows the tracking error e_h. From Fig. 6.3, it can be seen that, after 20 s, tracking error e_4 rises temporarily when actuator failure occurs and returns to a small neighborhood of the origin again, which shows that the actuator fault-tolerant control works in the systems. Compared with Figs. 6.2 and 6.4 shows that the systems will be out of control without the actuator failure compensation. In addition, the filtering error

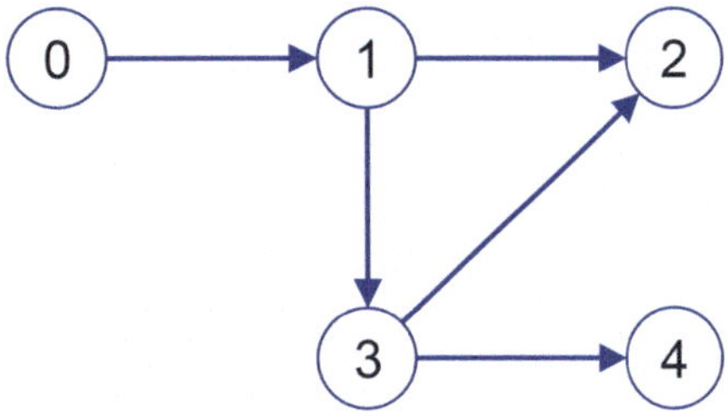

Fig. 6.1 Directed graph G

Table 6.1 Selection of Parameters ($h = 1, \ldots, 4$)

Initial conditions
$x_{h,1}(0) = 0$, $x_{h,2}(0) = 1$, $\hat{\theta}_{h,1}(0) = \hat{\theta}_{h,2}(0) = 0.1$, $\varrho_{h,10}(0) = \varrho_{h,11}(0) = 0$, $\hat{p}_{h,2}(0) = -0.5$, $\phi_h(0) = 4$
Controller parameters ($s = 1, 2$)
$k_{h,s1} = k_{h,s2} = 2$, $\gamma_{h,2} = 50$, $\tau_{h,21} = \tau_{h,22} = 0.01$, $\varepsilon_{h,1} = 0.01$, $\eta_{h,s} = 5$, $\sigma_{h,s1} = \sigma_{h,s2} = 0.05$, $\varsigma = 0.01$, $\mu = \frac{9}{7}$, $\beta = \frac{9}{11}$

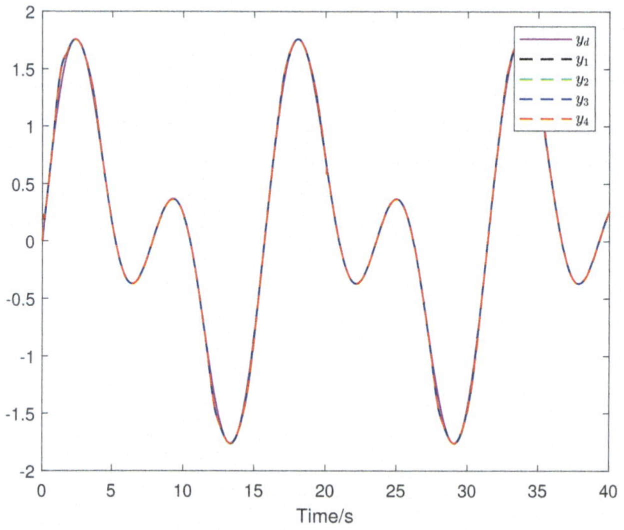

Fig. 6.2 The trajectories of followers y_h ($1 \le h \le 4$) and the leader y_d

compensation adaptive law $\hat{p}_{h,2}$ is shown in Fig. 6.5, the adaptive laws $\hat{\theta}_{h,j}$ ($j = 1, 2$) are shown in Fig. 6.6, and the control input signal u_h is shown in Fig. 6.7. With $\hat{p}_{h,2}$, $\hat{\theta}_{h,j}$ and u_h, the fixed-time consensus of MASs (6.1) is achieved.

Example 6.2 A practical example is shown to further illustrate the proposed control method. Consider the motion of i-th marine surface vehicle, it has the following nonlinear Norrbin dynamical model with stable line-movement

$$\begin{aligned} \dot{W}_i &= \nu_i \\ \dot{\nu}_i &= -\frac{1}{T_i}\nu_i - \frac{N}{T_i}\nu_i + \frac{\tau_i}{T_i}r_i \end{aligned}$$

where $\dot{W}_i$ is the course angle, ν_i and r_i show the course angle rate and command rudder angle with $i = 1, 2, 3, 4$. N represents the Norrbin coefficient. T_i and τ_i are defined as time and gain constants, respectively. Then, by defining and choosing parameters, one has

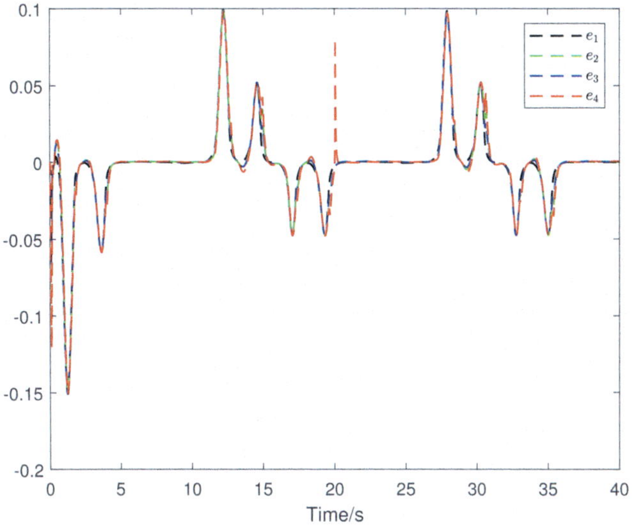

Fig. 6.3 The tracking errors of followers e_h $(1 \leq h \leq 4)$

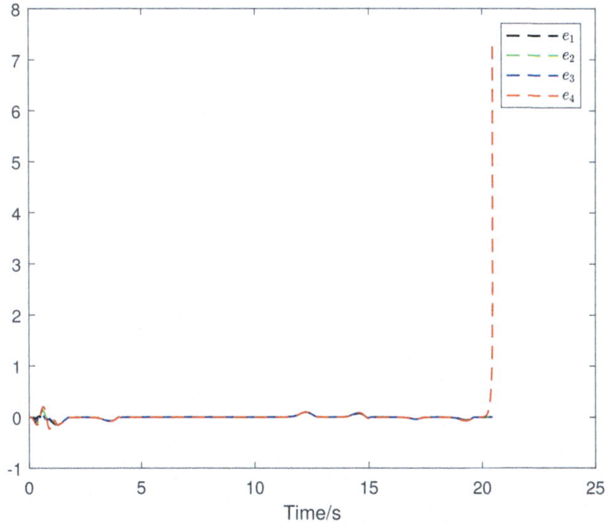

Fig. 6.4 The tracking errors of followers e_h $(1 \leq h \leq 4)$ without actuator failure compensation

$$\begin{aligned}
\dot{x}_{h,1} &= x_{h,2} \\
\dot{x}_{h,2} &= -\frac{x_{h,2}}{21} - \frac{0.3x_{h,2}^3}{21} + \frac{0.23u_{hf}}{21} \\
y_h &= x_{h,1}
\end{aligned}$$

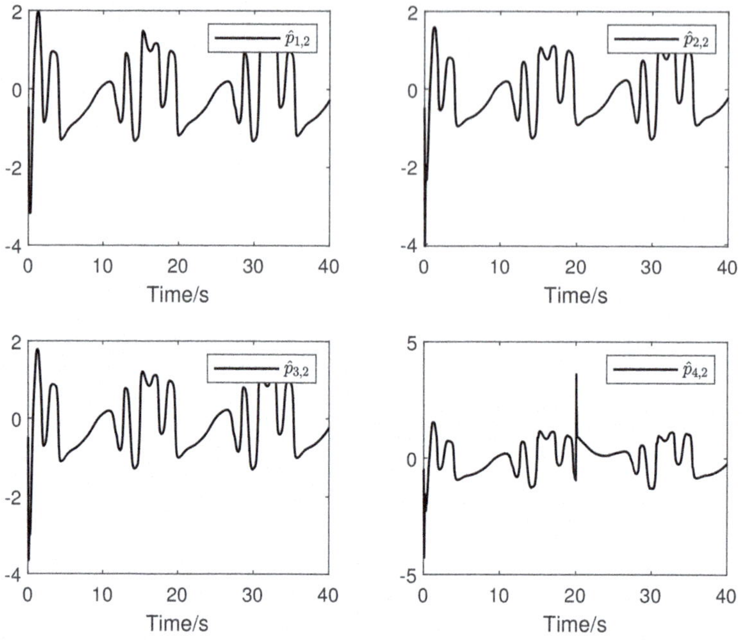

Fig. 6.5 The trajectories of filtering error compensation adaptive parameters $\hat{p}_{h,2}$ $(1 \leq h \leq 4)$

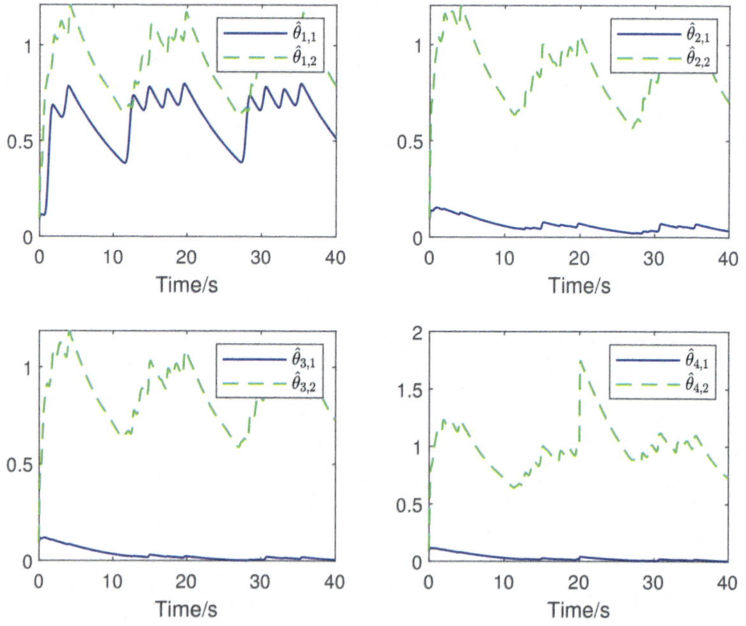

Fig. 6.6 The trajectories of adaptive parameters $\hat{\theta}_{h,j}$ $(1 \leq h \leq 4,\ j = 1, 2)$

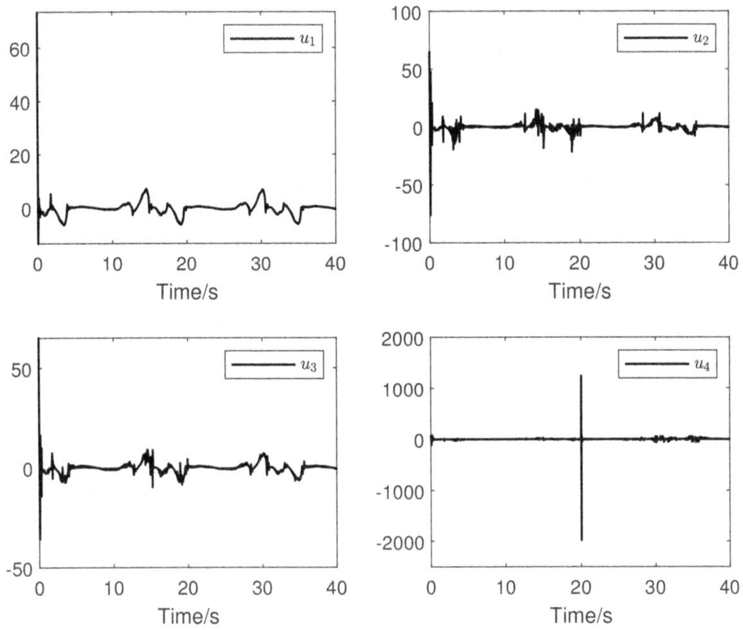

Fig. 6.7 The trajectories of control inputs u_h $(1 \le h \le 4)$

Table 6.2 Selection of Norrbin model parameters

Initial conditions
$x_{h,1}(0) = x_{h,2}(0) = 0$, $\hat{\theta}_{h,1}(0) = \hat{\theta}_{h,2}(0) = 0.01$, $\varrho_{h,10}(0) = \varrho_{h,11}(0) = 0$, $\hat{p}_{h,2}(0) = 0$, $\phi_h(0) = 8$
Controller paremeters
$k_{h,s1} = 2$, $k_{h,s2} = 5$, $\gamma_{h,2} = 50$, $\tau_{h,21} = \tau_{h,22} = 0.01$, $\varepsilon_{h,1} = 0.01$, $\eta_{h,s} = 5$, $\sigma_{h,s1} = \sigma_{h,s2} = 0.05$, $\varsigma = 0.01$, $\mu = \frac{9}{7}$, $\beta = \frac{9}{11}$

where $u_{1f} = u_1$, $u_{2f} = u_2 + 0.2\cos(t)$, $u_{3f} = 0.9u_3$, and

$$u_{4f} = \begin{cases} u_4, & t < 20 \\ -0.8u_4 + 0.1\cos(t), & t \ge 20. \end{cases}$$

Define the trajectory of leader as $y_d = \sin(t)$, the filter parameters are $R_{h,10} = 20$, $R_{h,11} = 5$, and other parameters are shown in Table 6.2.

The algorithm results based on this model are shown in Figs. 6.8 and 6.9. From Fig. 6.8, with the filtering error compensation mechanism and actuator fault-tolerant control mechanism, the output signals can track the reference signal even if actuator failures occur, and all signals are bounded in fixed time. In addition, when considering

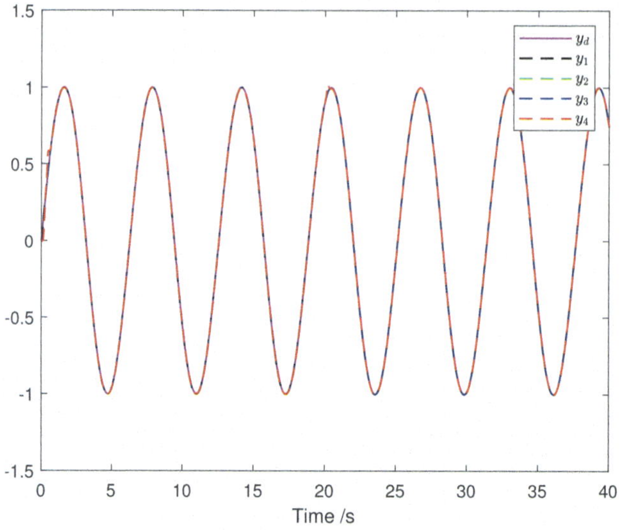

Fig. 6.8 The trajectories of followers y_h $(1 \leq h \leq 4)$ and the leader y_d

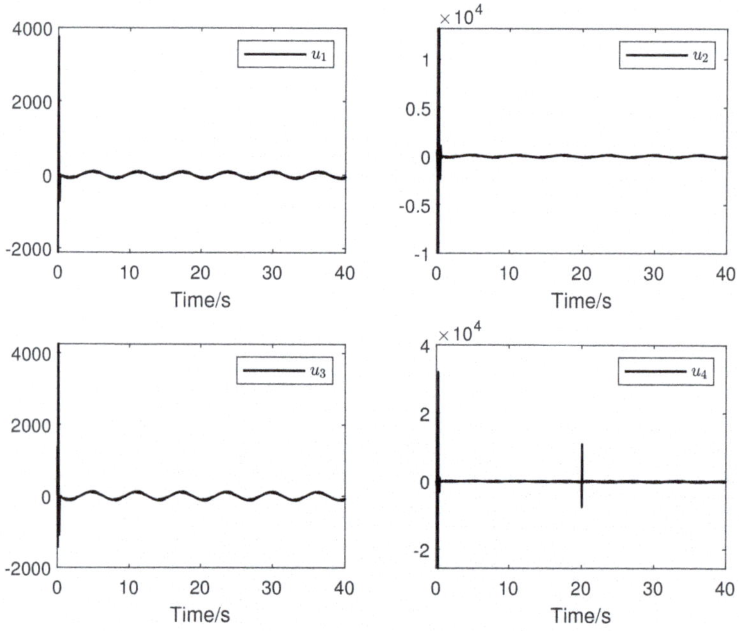

Fig. 6.9 The trajectories of control inputs u_h $(1 \leq h \leq 4)$

the communication constraints and time delay, the consensus problem of MASs will become more challenging (Zeng et al. 2022; Shangguan et al. 2021).

6.5 Conclusions

The fixed-time consensus problem of nonlinear MASs with actuator failures has been studied in this chapter. RBFNNs have been used to approximate the unknown functions. Command filtering has been utilized to solve the computation complexity problem. Furthermore, a novel filtering error compensation mechanism has been proposed to make filtering errors converge in fixed time. To avoid the effects of actuator failures, an actuator fault-tolerant control method based on Nussbaum gain theory has been utilized. Finally, the designed adaptive RBFNNs fixed-time command filtering controller guarantees that all signals are fixed-time bounded, and the effectiveness of the proposed control method has been verified by a numerical example and a practical example. In future work, the optimization algorithm for MASs will be considered to further improve control performance (Li et al. 2022; Gao et al. 2021). Furthermore, human-in-the-loop control is also an important method to improve the control scheme of MASs. For example, a novel human-in-the-loop control framework was developed, which was a major breakthrough and had an important reference value (Li et al. 2021, 2020a, b).

References

Cao L, Yao D, Li H, Meng W, Lu R (2023) Fuzzy-based dynamic event triggering formation control for nonstrict-feedback nonlinear MASs. Fuzzy Sets Syst 452:1–22

Chen M, Ge SS (2015) Adaptive neural output feedback control of uncertain nonlinear systems with unknown hysteresis using disturbance observer. IEEE Trans Indus Electron 62(12):7706–7716

Chen B, Zhang H, Liu X, Lin C (2018) Neural observer and adaptive neural control design for a class of nonlinear systems. IEEE Trans Neural Netw Learn Syst 29(9):4261–4271

Dai SL, He S, Chen X, Jin X (2020) Adaptive leader-follower formation control of nonholonomic mobile robots with prescribed transient and steady-state performance. IEEE Trans Indus Inf 16(6):3662–3671

Dai SL, Lu K, Fu J (2022) Adaptive finite-time tracking control of nonholonomic multirobot formation systems with limited field-of-view sensors. IEEE Trans Cybernet 52(10):10695–10708

Deng C, Gao W, Che W (2020) Distributed adaptive fault-tolerant output regulation of heterogeneous multi-agent systems with coupling uncertainties and actuator faults. IEEE/CAA J Automatica Sinica 7(4):1098–1106

Gao H, Li Z, Yu X, Qiu J (2021) Hierarchical multiobjective heuristic for PCB assembly optimization in a beam-head surface mounter. IEEE Trans Cybernet. https://doi.org/10.1109/TCYB.2020.3040788

Jin Z, Qin Z, Zhang X, Guan C (2022) A leader-following consensus problem via a distributed observer and fuzzy input-to-output small-gain theorem. IEEE Trans Control Netw Syst 9(1):62–74

Jin Z, Ahn CK, Li J (2023) Momentum-based distributed continuous-time nonconvex optimization of nonlinear multi-agent systems via timescale separation. IEEE Trans Netw Sci Eng 10(2):980–989

Jin Z, Sun X, Qin Z, Ahn CK (2022) Fuzzy small-gain approach for the distributed optimization of T-S fuzzy cyber-physical systems. IEEE Trans Cybernet. https://doi.org/10.1109/TCYB.2022.3202576

Li Y, Tong S (2017) Adaptive neural networks decentralized FTC design for nonstrict-feedback nonlinear interconnected large-scale systems against actuator faults. IEEE Trans Neural Netw Learn Syst 28(11):2541–2554

Li K, Tong S, Li Y (2020) Finite-time adaptive fuzzy decentralized control for nonstrict-feedback nonlinear systems with output-constraint. IEEE Trans Syst Man Cybernet Syst 50(12):5271–5284
Li Z, Deng C, Zhao K (2020a) Human-cooperative control of a wearable walking exoskeleton for enhancing climbing stair activities. IEEE Trans Indus Electron 67(4):3086–3095
Li Z, Ren Z, Zhao K, Deng C, Feng Y (2020b) Human-cooperative control design of a walking exoskeleton for body weight support. IEEE Trans Indus Inf 16(5):2985–2996
Li Y, Qu F, Tong S (2021) Observer-based fuzzy adaptive finite-time containment control of nonlinear multiagent systems with input delay. IEEE Trans Cybernet 51(1):126–137
Li Z, Zhao K, Zhang L, Wu X, Zhang T, Li Q, Li X, Su CY (2021) Human-in-the-loop control of a wearable lower limb exoskeleton for stable dynamic walking. IEEE/ASME Trans Mechatron 26(5):2700–2711
Li Z, Yu X, Qiu J, Gao H (2022) Cell division genetic algorithm for component allocation optimization in multifunctional placers. IEEE Trans Indus Inf 18(1):559–570
Li H, Wu Y, Chen M, Lu R (2023) Adaptive multigradient recursive reinforcement learning event-triggered tracking control for multiagent systems. IEEE Trans Neural Netw Learn Syst 34(1):144–156
Li Z, Gao L, Chen W, Xu Y (2020) Distributed adaptive cooperative tracking of uncertain nonlinear fractional-order multi-agent systems. IEEE/CAA J Automatica Sinica 7(1):292–300
Lin X, Yu Y, Sun C (2019) A decoupling control for quadrotor UAV using dynamic surface control and sliding mode disturbance observer. Nonlinear Dyn 97(1):781–795
Lin G, Li H, Ahn CK, Yao D (2022) Event-based finite-time neural control for human-in-the-loop UAV attitude systems. IEEE Trans Neural Netw Learn Syst. https://doi.org/10.1109/TNNLS.2022.3166531
Liu Z, Wang F, Zhang Y, Chen X, Chen CLP (2015) Adaptive tracking control for a class of nonlinear systems with a fuzzy dead-zone input. IEEE Trans Fuzzy Syst 23(1):193–204
Liu Z, Dong X, Xue J, Li H, Chen Y (2016) Adaptive neural control for a class of pure-feedback nonlinear systems via dynamic surface technique. IEEE Trans Neural Netw Learn Syst 27(9):1969–1975
Liu YJ, Li J, Tong S, Chen CLP (2016) Neural network control-based adaptive learning design for nonlinear systems with full-state constraints. IEEE Trans Neural Netw Learn Syst 27(7):1562–1571
Liu J, Zhang Y, Yu Y, Sun C (2019) Fixed-time event-triggered consensus for nonlinear multiagent systems without continuous communications. IEEE Trans Syst Man Cybernet Syst 49(11):2221–2229
Liu Y, Li H, Zuo Z, Li X, Lu R (2022) An overview of finite/fixed-time control and its application in engineering systems. IEEE/CAA J Automatica Sinica 9(12):2106–2120
Liu Z, Lin W, Yu X, Rodríguez-Andina JJ, Gao H (2022) Approximation-free robust synchronization control for dual-linear-motors-driven systems with uncertainties and disturbances. IEEE Trans Indus Electron 69(10):10500–10509
Liu T, Qin Z, Hong Y, Jiang ZP (2022) Distributed optimization of nonlinear multiagent systems: a small-gain approach. IEEE Trans Autom Control 67(2):676–691
Ma H, Ren H, Zhou Q, Li H, Wang Z (2022) Observer-based neural control of N-link flexible-joint robots. IEEE Trans Neural Netw Learn Syst. https://doi.org/10.1109/TNNLS.2022.3203074
Ni J, Shi P, Zhao Y, Wu Z (2021) Fixed-time output consensus tracking for high-order multi-agent systems with directed network topology and packet dropout. IEEE/CAA J Automatica Sinica 8(4):817–836
Pan Y, Li Q, Liang H, Lam HK (2022) A novel mixed control approach for fuzzy systems via membership functions online learning policy. IEEE Trans Fuzzy Syst 30(9):3812–3822
Pan Y, Wu Y, Lam HK (2022) Security-based fuzzy control for nonlinear networked control systems with DoS attacks via a resilient event-triggered scheme. IEEE Trans Fuzzy Syst 30(10):4359–4368
Polyakov A, Efimov D, Perruquetti W (2015) Finite-time and fixed-time stabilization: implicit Lyapunov function approach. Automatica 51:332–340

Ren H, Wang Y, Liu M, Li H (2022) An optimal estimation framework of multi-agent systems with random transport protocol. IEEE Trans Signal Process 70:2548–2559

Ren H, Ma H, Li H, Lu R (2023) A disturbance observer based intelligent control for nonstrict-feedback nonlinear systems. Sci China-Technol Sci 66(2):456–467

Shangguan XC, Zhang CK, He Y, Jin L, Jiang L, Spencer JW, Wu M (2021) Robust load frequency control for power system considering transmission delay and sampling period. IEEE Trans Indus Inf 17(8):5292–5303

Wang L, Chen CLP (2020) Adaptive fuzzy dynamic surface control of nonlinear constrained systems with unknown virtual control coefficients. IEEE Trans Fuzzy Syst 28(8):1737–1747

Wang Y, Song Y, Krstic M, Wen C (2016) Adaptive finite time coordinated consensus for high-order multi-agent systems: adjustable fraction power feedback approach. Inf Sci 372:392–406

Wang F, Chen B, Sun Y, Gao Y, Lin C (2020) Finite-time fuzzy control of stochastic nonlinear systems. IEEE Trans Cybernet 50(6):2617–2626

Wang L, Wang H, Liu PX, Ling S, Liu S (2022) Fuzzy finite-time command filtering output feedback control of nonlinear systems. IEEE Trans Fuzzy Syst 30(1):97–107

Wei Q, Wang X, Zhong X, Wu N (2021) Consensus control of leader-following multi-agent systems in directed topology with heterogeneous disturbances. IEEE/CAA J Automatica Sinica 8(2):423–431

Xia J, Zhang J, Feng J, Wang Z, Zhuang G (2021) Command filter-based adaptive fuzzy control for nonlinear systems with unknown control directions. IEEE Trans Syst Man Cybernet Syst 51(3):1945–1953

Zeng HB, He Y, Teo KL (2022) Monotone-delay-interval-based lyapunov functionals for stability analysis of systems with a periodically varying delay. Automatica 138:110030

Zhang JX, Chai T (2022) Singularity-free continuous adaptive control of uncertain underactuated surface vessels with prescribed performance. IEEE Trans Syst Man Cybernet Syst 52(9):5646–5655

Zhang L, Chen B, Lin C, Shang Y (2022) Fuzzy adaptive fixed-time consensus tracking control of high-order multi-agent systems. IEEE Trans Fuzzy Syst 30(2):567–578

Zhang JX, Wang QG, Ding W (2022) Global output-feedback prescribed performance control of nonlinear systems with unknown virtual control coefficients. IEEE Trans Automat Control 67(12):6904–6911

Zhang P, Liu T, Jiang ZP (2023) Tracking control of unicycle mobile robots with event-triggered and self-triggered feedback. IEEE Trans Automat Control 68(4):2261–2276

Zheng X, Li H, Ahn CK, Yao D (2023) NN-based fixed-time attitude tracking control for multiple unmanned aerial vehicles with nonlinear faults. IEEE Trans Aerosp Electron Syst 59(2):1738–1748

Zhou H, Sui S, Tong S (2022) Fuzzy adaptive finite-time consensus control for high-order nonlinear multi-agent systems based on event-triggered. IEEE Trans Fuzzy Syst 30(11):4891–4904

Chapter 7
Event-Triggered Adaptive Fault-Tolerant Control with Unknown Disturbances

Abstract This chapter presents an event-triggered consensus control protocol for a class of multi-agent systems with actuator faults, sensor faults, and unknown disturbances. The adaptive neural network compensation control method is introduced to solve the problem of sensor faults. The event-triggered mechanism is developed to reduce the communication burden. In the control design process, the radial basis function neural networks are used to approximate the unknown nonlinear functions, and a nonlinear disturbance observer is used to eliminate the effect of unknown external disturbances. Furthermore, based on the graph theory and Lyapunov stability theory, it is further shown that the consensus tracking errors are semi-globally uniformly ultimately bounded. Finally, the simulation example illustrates the effectiveness of the designed control protocol.

7.1 Introduction

Over the last decade, some control problems have been investigated such as finite-time control (Xie et al. 2021; Li et al. 2019; Du et al. 2021; Li et al. 2019; Liu et al. 2018; Zhou et al. 2021c), delay-dependent control (Li and Wu 2016; Liu et al. 2020; Li and Song 2016), networked control (Lu et al. 2014; Zhang et al. 2019; Lu et al. 2012). However, the cooperative tracking control problems of multi-agent systems (MASs) have received much more attention (Hu et al. 2016; Liu et al. 2019; Wang et al. 2020; Liang et al. 2021; Su et al. 2019; Zhou et al. 2021a, b). As far as we know, the cooperative tracking control problems of MASs widely exist in the practical applications, such as spacecraft formation (Ren 2010), animal groups (Sumpter and Pratt 2003) and distributed sensor networks (Saber and Murray 2003). The consensus problem of MASs is an important topic and the distributed cooperative control problem has been intensively investigated in the past few years, e.g., Hou et al. (2009), Hu et al. (2018), Hua et al. (2018), Zhang et al. (2018), Wu and Su (2018). In Hou et al. (2009), the consensus control problem of MASs with uncertain nonlinear dynamics and external disturbances was addressed, and a robust adaptive control approach was proposed to deal with the approximation errors and external disturbances. Hu et al. (2018) considered the consensus control problem of high-order

H. Ren et al., *Intelligent Control of Nonlinear Systems and Nonlinear Multi-Agent Systems with Complex Constraints*, Intelligent Control and Learning Systems 16,
https://doi.org/10.1007/978-981-96-0320-6_7

MASs with antagonistic interactions, and a novel control strategy was developed by using the relative state information from its neighbors. The leader-following output consensus control problem for a class of high-order nonlinear MASs was studied in Hua et al. (2018). Zhang et al. (2018) investigated the consensus tracking control problem of nonlinear MASs with state constraints and unknown disturbances. Wu and Su (2018) studied the consensus control problem for positive MASs with nonlinear control input, and the global and local consensus control results were analyzed for the case of saturation-type sector input nonlinearities.

Compared with the time-sampling, the event-triggered control strategy can reduce resource consumption and improve resource utilization (Yao et al. 2020; Leng et al. 2019; Zhang et al. 2021; Hui et al. 2020; Chen et al. 2020; Zhu et al. 2020, 2021). Hence, the event-triggered control of MASs has attracted considerable attention. More specifically, Tan et al. (2018) investigated the leader-following consensus control problem of MASs by using a distributed event-triggered impulsive control method. You et al. (2018) addressed the distributed event-triggered consensus control problem for a class of nonlinear MASs subject to actuator saturation. The output consensus control problem was considered for heterogeneous linear MASs via event-triggered control method in Qian et al. (2018). The event-triggered consensus control problem was studied in Ye et al. (2018) for the fractional-order MASs. Wang et al. (2018) investigated the tracking consensus control problem for the second-order leader MASs by designing fractional-order follower observer, where a periodic sampled-based data event-triggered control method was employed. In addition, an event-triggered strategy was presented for MASs to solve the cooperative output regulation problem under switching communication topologies in Hu et al. (2018). Zheng et al. (2018) studied the event-triggered problem for interconnected switched MASs.

From another point of view, some different fault-tolerant consensus control methods have been studied for MASs. In Zhang and Yang (2018), the output feedback control problem was considered for a class of uncertain nonlinear systems with actuator failures. The problem of consensus tracking control was addressed for MASs with actuator faults under directed networks in Wu et al. (2018). Luo and Ye (2019) considered the event-triggered consensus control problem for linear MASs with actuator faults. The work in Sakthivel et al. (2019) addressed the consensus fault-tolerant control problem for MASs subject to linear fractional transformation uncertain parameters. In Sakthivel et al. (2018), the robust fault-tolerant control protocol with actuator saturation was proposed to solve the consensus control problem of MASs. The event-triggered control method for MASs with sensor faults was proposed to solve the consensus control problem (Cao et al. 2021).

Inspired by the above observations, the event-triggered control problem for MASs with unknown disturbances, sensor, and actuator faults are considered in this chapter. The main contributions and difficulties of this chapter are highlighted as follows:

1. In order to eliminate the effect of unknown external disturbances, the disturbance observer is designed in this chapter, and the unknown nonlinear functions are considered in MASs with unknown disturbances, which is a new challenge for the design of the disturbance observer.

2. Different from the works in Luo and Ye (2019), Sakthivel et al. (2019), Sakthivel et al. (2018), the fault problems are considered for the MASs in this chapter, which not only contains actuator faults but also includes sensor faults. In addition, the sensor faults problem is solved via the proposed adaptive neural network (NN) compensation control method.
3. An event-triggered fault-tolerant consensus control protocol is provided for MASs. Compared with the results in Ye et al. (2018), Wang et al. (2016) that only deal with a limited number of actuator faults, the event-triggered control mechanism is introduced in this chapter for MASs with an infinite number of actuator failures, and the communication resources can also be successfully saved.

7.2 Preliminaries

7.2.1 *Basic Graph Theory*

The information exchanges among the agents can be modeled by a directed graph $\zeta = (\mathcal{V}, \mathcal{E}, \mathcal{A})$ including a set of agents $\mathcal{V} = \{1, \ldots, N\}$, a set of edges $\mathcal{E} \subseteq \mathcal{V} \times \mathcal{V}$ and the relevant adjacency matrix $\mathcal{A} = \left[a_{i,j}\right] \in \mathbb{R}^{N\times N}$, which is defined as $a_{i,i} = 0$, $a_{i,j} > 0$ if an edge $(\mathcal{V}_i, \mathcal{V}_j) \in \mathcal{E}$ and $a_{i,j} = 0$, otherwise. The Laplacian matrix $L = (l_{i,j}) = D - A$ and $L \times \mathbf{1}_N = 0$, where $\mathbf{1}_N$ is a column vector with 1 for each element, $D = \mathrm{diag}(d_1, \ldots, d_N)$ and the in-degree for agent i is $d_i = \sum_{j=1}^{N} a_{i,j}$. The $\zeta_\vdash = (\mathcal{V}_\vdash, \mathcal{E}_\vdash)$ is augmented graph with $\mathcal{V}_\vdash = \{0, 1, \ldots, N\}$ and $\mathcal{E}_\vdash \subseteq \mathcal{V}_\vdash \times \mathcal{V}_\vdash$, where 0 is called as leader. The directed graph ζ is defined as a spanning tree, if there exists at least a directed path from the root to all other agents.

Lemma 7.1 (Zhang et al. 2011) *If denote $B = diag(a_{1,0}, \ldots, a_{N,0})$, all eigenvalues of the matrix $L + B$ have positive real part if and only if the graph contains a spanning tree with vertex 0 as the root.*

7.2.2 *Problem Formulation*

For $i = 1, \ldots, N$, the system model of the ith agent is described by

$$
\begin{cases}
\dot{x}_{i,m} = x_{i,m+1} + f_{i,m}(x_i) + H_{i,m}(t) \\
\dot{x}_{i,n} = \sum_{p=1}^{M} b_{i,p}u_{i,p} + f_{i,n}(x_i) + H_{i,n}(t) \\
y_i = h_i(x_{i,1})
\end{cases}
\tag{7.1}
$$

where $x_i = \left[x_{i,1}, \ldots, x_{i,n}\right]^T \in \mathbb{R}^n$, $u_i = [u_{i,1}, u_{i,2}, \ldots, u_{i,M}]^T \in \mathbb{R}^M$ and y_i stand for the system state, control input, and the system output, respectively. For $m = 1, 2, \ldots, n-1$, $f_{i,m}(x_i)$ and $f_{i,n}(x_i)$ are unknown nonlinear functions, $H_{i,m}(t)$ and $H_{i,n}(t)$ are unknown external disturbances, $b_{i,p}$ is known scalar, and $h_i(x_{i,1}) = k_i(\cdot)x_{i,1} + \rho_i(\cdot)$, where $k_i(\cdot)$ and $\rho_i(\cdot)$ denote the parameters of sensor faults.

Remark 7.1 The actuator fault model is considered in this chapter. Therefore, the controller design procedure is more difficult than the ordinary tracking control problem. Moreover, the sensor faults are considered in each agent output, which is a challenge to solve faults problem.

The actuator fault model (Li and Yang 2016) is described by

$$\begin{aligned} u_{i,p}^F(t) &= k_{i,p,g}u_{i,p}(t) + \bar{u}_{i,p,g}(t), t \in [t_{i,p,g}^S, t_{i,p,g}^E] \\ k_{i,p,g}\bar{u}_{i,p,g}(t) &= 0 \end{aligned}$$

where $k_{i,p,g} \in [0, 1]$, $t_{i,p,g}^S$, $t_{i,p,g}^E$ are all unknown constants, and $\bar{u}_{i,p,g}(t)$ is an unknown, bounded and piecewise continuous signal, where $p = 1, 2, \ldots, M$, $g = 1, 2, \ldots$.

Define

$$k_{i,p}(t) = \begin{cases} k_{i,p,g} & \text{if } t \in [t_{i,p,g}^S, t_{i,p,g}^E] \\ 1 & \text{if } t \in (t_{i,j,g}^E, t_{i,j,g+1}^S] \end{cases}$$

$$\breve{u}_{i,p}(t) = \begin{cases} \bar{u}_{i,p,g}(t) & \text{if } t \in [t_{i,p,g}^S, t_{i,p,g}^E] \\ 0 & \text{if } t \in (t_{i,p,g}^E, t_{i,p,g+1}^S] \end{cases}.$$

The actuator faults model can be expressed as

$$u_{i,p}^F(t) = k_{i,p}(t)u_{i,p}(t) + \breve{u}_{i,p}(t).$$

Assumption 7.1 There exists an unknown constant $\bar{u}_{i,p}$ such that $0 < \breve{u}_{i,p}(t) \le \bar{u}_{i,p}$.

Assumption 7.2 For the ith agent, up to $M - 1$ actuator faults are allowed at the same time, and the desired control objective can still be achieved by the remaining actuators.

In sensor fault model (Bounemeur et al. 2018), let $f_{si} = (k_i(\cdot) - 1)x_{i,1} + \rho_i(\cdot)$. Therefore, y_i can be rewritten as

$$y_i = x_{i,1} + f_{si}.$$

The time derivative of y_i is presented as

$$\dot{y}_i = \dot{x}_{i,1} + f_{psi}$$

where $f_{psi} = \dot{f}_{si}$.

7.2.3 Radial Basis Function Neural Networks

The following radial basis function neural networks (RBF NNs) are used to estimate nonlinear functions (Liu et al. 2020):

$$\bar{f}(Z) = W^{*T} S(Z) + F(Z), \quad \forall Z \in \Omega \subset \mathbb{R}^m, |F(Z)| \le \varepsilon$$

where $F(Z)$ is the approximation error. $S(Z) = [S_1(Z), S_2(Z), \ldots, S_k(Z)]^T$. Furthermore, $S_i(Z)$ denotes the Gaussian basis function

$$S_i(Z) = \exp\left[-\frac{(Z-\iota_i)^T (Z-\iota_i)}{\upsilon_i^2}\right]$$

where $\iota_i = \left[\iota_{i,1}, \iota_{i,2}, \ldots, \iota_{i,q}\right]^T$ is the receptive field's center and υ_i represents the width of the Gaussian basis function with $i = 1, \ldots, k$. Additionally, $W^* = [W_1, W_2, \ldots, W_k]^T \in \mathbb{R}^k$ is weight vector which defined as

$$W^* = \arg \min_{W \in \mathbb{R}^k} \left\{ \sup_{Z \in \Omega} \left|\bar{f}(Z) - W^T S(Z)\right| \right\}$$

where $W \in \mathbb{R}^k$.

Lemma 7.2 (Su et al. 2016) *Choose* $S(\bar{x}_c) = [S_1(\bar{x}_c), S_2(\bar{x}_c), \ldots, S_k(\bar{x}_c)]^T$ *where* $\bar{x}_c = [x_1, \ldots, x_c]^T$ *is the RBF NNs' basis function vector. For any positive integer* $c \ge p$, *the following inequality holds*

$$\|S(\bar{x}_c)\|^2 \le \left\|S\left(\bar{x}_p\right)\right\|^2$$

7.2.4 Disturbance Observer Design

The unknown nonlinear function $f_{i,k}(x_i)$ is estimated by RBF NNs as follows

$$f_{i,k}(x_i) = M_{i,k}^{*} S_{i,k}(x_i) + \upsilon_{i,k}(x_i), \quad \forall x_i \in \Omega_{x_i}$$

where $M_{i,k}^*$ means the weight vector, $S_{i,k}(x_i)$ denotes the Gaussian basis function, and $\upsilon_{i,k}(x_i)$ is approximation error satisfying $\left|\upsilon_{i,k}(x_i)\right| \le \upsilon_{i,k,0}$ with $\upsilon_{i,k,0} > 0$. Then, we can get

$$\begin{cases} \dot{x}_{i,m} = x_{i,m+1} + M_{i,m}^{*T} S_{i,m}(x_i) + \upsilon_{i,m}(x_i) + H_{i,m}(t) \\ \dot{x}_{i,n} = \sum_{p=1}^{M} b_{i,p}(k_{i,p}u_{i,p} + \bar{u}_{i,p}) + M_{i,n}^{*T} S_{i,n}(x_i) + \upsilon_{i,n}(x_i) + H_{i,n}(t) \\ y_i = h_i(x_{i,1}). \end{cases}$$

The disturbance observer is designed as follows

$$\begin{aligned} \hat{z}_{i,k} &= \lambda_{i,k}(x_{i,k} - Q_{i,k}) \\ \dot{Q}_{i,k} &= x_{i,k+1} + \hat{z}_{i,k} + \hat{M}_{i,k}^{T} S_{i,k}(x_i) \end{aligned}$$

where $\lambda_{i,k} > 0$, $\tilde{M}_{i,k} = M_{i,k}^{*} - \hat{M}_{i,k}$ and $x_{i,n+1} = u_i$. Besides, we can separately get the disturbance estimation and the disturbance estimation error as follows

$$\begin{aligned} \dot{\hat{z}}_{i,k} &= \lambda_{i,k}(\dot{x}_{i,k} - \dot{Q}_{i,k}) \\ &= \lambda_{i,k}\left(H_{i,k}(t) - \hat{z}_{i,k} + \tilde{M}_{i,k}^{T} S_{i,k}(x_i) + \upsilon_{i,k}(x_i)\right) \\ \dot{\tilde{z}}_{i,k} &= -\lambda_{i,k}\tilde{z}_{i,k} + \dot{H}_{i,k}(t) - \lambda_{i,k}(\tilde{M}_{i,k} S_{i,k}(x_i) + \upsilon_{i,k}(x_i)) \\ \tilde{z}_{i,k} &= H_{i,k}(t) - \hat{z}_{i,k}. \end{aligned}$$

7.3 Control Law Design and Stability Analysis

The tracking errors are presented as

$$\begin{aligned} e_{i,1} &= \sum_{j=1}^{N} a_{i,j}(y_i - y_j) + a_{i,0}(y_i - y_r) \\ e_{i,\beta} &= x_{i,\beta} - \alpha_{i,\beta-1} \end{aligned}$$

where $i = 1, \ldots, N$, $\beta = 2, \ldots, n$. $\alpha_{i,\beta-1}$ is the virtual controller, and $a_{i,0} \geq 0$. Define

$$\theta_i = \max\left\{\left\| W_{i,k}^{*}, k = 1, \ldots, n \right\|, i = 1, \ldots, N\right\}$$

where $\tilde{\theta}_i = \theta_i - \hat{\theta}_i$.

Lemma 7.3 (Zhang and Lewis 2012) *Define* $e_1 = col\{e_{1,1}, e_{2,1}, \ldots, e_{N,1}\}$, $y = col\{y_1, y_2, \ldots, y_N\}$, $\bar{y}_r = \mathbf{1}_N \otimes y_r$, *then*

$$\|y - \bar{y}_r\| \leq \frac{\|e_1\|}{\bar{\sigma}((L + B))}$$

where $\bar{\sigma}(L + B)$ is minimum singular value.

Step 1: The derivative of $e_{i,1}$ yields

$$\dot{e}_{i,1} = (d_i + a_{i,0})(x_{i,2} + f_{i,1}(x_i) + H_{i,1}(t) + f_{psi}) - \sum_{j\in N_i} a_{i,j}(x_{j,2} + f_{j,1}(x_j) + f_{psj} + H_{j,1}(t)) - a_{i,0}\dot{y}_r.$$

Consider the Lyapunov candidate as

$$V_{i,1} = \frac{1}{2}e_{i,1}^2 + \frac{1}{2\gamma_i}\tilde{\theta}_i^2 + \frac{\tilde{z}_{i,1}^2}{2} + \frac{1}{2\eta_{i,1}}\tilde{M}_{i,1}^T\tilde{M}_{i,1}$$

where $\gamma_i > 0$ and $\eta_{i,1} > 0$. Then, we get

$$\begin{aligned}
\dot{V}_{i,1} &= e_{i,1}\left[(d_i + a_{i,0})(x_{i,2} + f_{i,1}(x_i) + H_{i,1}(t) + f_{psi}) - a_{i,0}\dot{y}_r \right.\\
&\quad \left. - \sum_{j\in N_i} a_{i,j}(x_{j,2} + f_{j,1}(x_j) + f_{psj} + H_{j,1}(t))\right] - \frac{1}{\eta_{i,1}}\tilde{M}_{i,1}^T\dot{\hat{M}}_{i,1}\\
&\quad + \tilde{z}_{i,1}(-\lambda_{i,1}\tilde{z}_{i,1} + \dot{H}_{i,1} - \lambda_{i,1}\tilde{M}_{i,1}S_{i,1}(x_i) - \lambda_{i,1}\upsilon_{i,1}(x_i)) - \frac{1}{\gamma_i}\tilde{\theta}_i\dot{\hat{\theta}}_i\\
&\le e_{i,1}[(d_i + a_{i,0})(e_{i,2} + \alpha_{i,1}) + \bar{f}_{i,1}(Z_{i,1}) - (d_i + a_{i,0})\tilde{z}_{i,1}\\
&\quad - \sum_{j\in N_i} a_{i,j}\tilde{z}_{j,1}] + \frac{d_i}{2}e_{i,1}^2 - \frac{d_i + a_{i,0}}{2}e_{i,1}^2 - \lambda_{i,1}\tilde{z}_{i,1}^2 + \tilde{z}_{i,1}\dot{H}_{i,1}\\
&\quad - \frac{1}{\eta_{i,1}}\tilde{M}_{i,1}^T(\dot{\hat{M}}_{i,1} + \eta_{i,1}\lambda_{i,1}\tilde{z}_{i,1}S_{i,1}(x_i))\\
&\quad - \lambda_{i,1}\tilde{z}_{i,1}\upsilon_{i,1}(x_i) - \frac{1}{\gamma_i}\tilde{\theta}_i\dot{\hat{\theta}}_i
\end{aligned} \tag{7.2}$$

with

$$\begin{aligned}
\bar{f}_{i,1}(Z_{i,1}) &= (d_i + a_{i,0})(f_{i,1}(x_j) + H_{i,1}(t) + f_{psi}) - \sum_{j\in N_i} a_{i,j}(x_{j,2} + f_{j,1}(x_j)\\
&\quad + f_{psj} + H_{j,1}(t)) + \frac{d_i + a_{i,0}}{2}e_{i,1}^2 + \frac{d_i}{2}e_{i,1}^2 - a_{i,0}\dot{y}_r.
\end{aligned}$$

By using RBF NNs, the unknown nonlinear function $\bar{f}_{i,1}(Z_{i,1})$ is estimated as follows

$$\bar{f}_{i,1}(Z_{i,1}) = W_{i,1}^{*T}S_{i,1}(Z_{i,1}) + F_{i,1}(Z_{i,1})$$

where $Z_{i,1} = \left[x_i^T, x_j^T, y_0, \dot{y}_0, \hat{z}_{i,1}, \hat{z}_{j,1}\right]^T$, and $F_{i,1}(Z_{i,1})$ denotes the estimation error which satisfies $0 < |F_{i,1}(Z_{i,1})| \le \varepsilon_{i,1}$. Based on the Young's inequality and Lemma 7.2, one has

$$
\begin{aligned}
e_{i,1}\bar{f}_{i,1}(Z_{i,1}) &\leq \frac{\|W_{i,1}^*\|^2}{2p_{i,1}^2}e_{i,1}^2S_{i,1}^T(Z_{i,1})S_{i,1}(Z_{i,1}) + \frac{p_{i,1}^2}{2} + \frac{e_{i,1}^2}{2} + \frac{\varepsilon_{i,1}^2}{2} \\
&\leq \frac{\theta_{i,1}}{2p_{i,1}^2}e_{i,1}^2S_{i,1}^T(\bar{Z}_{i,1})S_{i,1}(\bar{Z}_{i,1}) + \frac{p_{i,1}^2}{2} + \frac{e_{i,1}^2}{2} + \frac{\varepsilon_{i,1}^2}{2}
\end{aligned}
$$

where $p_{i,1} > 0$ and $\bar{Z}_{i,1} = [x_{i,1}, x_{j,1}, y_0, \dot{y}_0, \hat{z}_{i,1}, \hat{z}_{j,1}]^T$. Then, it yields,

$$
\begin{aligned}
e_{i,1}(d_i + a_{i,0})\tilde{z}_{i,1} &\leq \frac{d_i + a_{i,0}}{2}e_{i,1}^2 + \frac{d_i + a_{i,0}}{2}\tilde{z}_{i,1}^2 \\
\sum_{j\in N_i} a_{i,j}\tilde{z}_{j,1}e_{i,1} &\leq \frac{d_i}{2}e_{i,1}^2 + \sum_{j\in N_i} a_{i,j}\frac{\tilde{z}_{j,1}^2}{2} \\
-\lambda_{i,1}\tilde{z}_{i,1}\upsilon_{i,1}(x_i) &\leq \frac{1}{2}\tilde{z}_{i,1}^2 + \frac{\lambda_{i,1}^2\upsilon_{i,1,0}^2}{2}.
\end{aligned}
$$

Define

$$
\begin{aligned}
\alpha_{i,1} &= \frac{1}{d_i + a_{i,0}}[-k_{i,1}e_{i,1} - \frac{e_{i,1}}{2} - \frac{\hat{\theta}_i}{2p_{i,1}^2}e_{i,1}S_{i,1}^T(\bar{Z}_{i,1})S_{i,1}(\bar{Z}_{i,1})] \\
\dot{\hat{M}}_{i,1} &= -\eta_{i,1}\lambda_{i,1}\hat{z}_{i,1}S_{i,1}(\bar{Z}_{i,1}) - l_{i,1}\hat{M}_{i,1}
\end{aligned}
$$

where $k_{i,1} > 0$ and $l_{i,1} > 0$. By using the above-mentioned RBF NNs, the Young's inequality, the virtual controller $\alpha_{i,1}$ and the adaptive law $\dot{\hat{M}}_{i,1}$, the following formula holds

$$
\begin{aligned}
\dot{V}_{i,1} \leq& -k_{i,1}e_{i,1}^2 + (d_i + a_{i,0})e_{i,1}e_{i,2} + \tilde{\theta}_i\left(\frac{1}{2p_{i,1}^2}e_{i,1}^2S_{i,1}^T(\bar{Z}_{i,1})S_{i,1}(\bar{Z}_{i,1}) - \frac{\dot{\hat{\theta}}_i}{r_i}\right) \\
&+\frac{(d_i + a_{i,0})}{2}\tilde{z}_{i,1}^2 + \frac{1}{2}\sum_{j\in N_i} a_{i,j}\tilde{z}_{j,1}^2 - \left(\lambda_{i,1} - \frac{1}{2}\right)\tilde{z}_{i,1}^2 + \tilde{z}_{i,1}\dot{H}_{i,1} \\
&+\frac{p_{i,1}^2}{2} + \frac{l_{i,1}}{\eta_{i,1}}\tilde{M}_{i,1}^T\hat{M}_{i,1} + \frac{\varepsilon_{i,1}^2}{2} + \frac{\lambda_{i,1}^2\upsilon_{i,1,0}^2}{2}.
\end{aligned}
$$

Step m (m = 2, ..., n − 1): The derivatives of error $e_{i,m}$ yields

$$
\dot{e}_{i,m} = x_{i,m+1} + f_{i,m}(x_i) - \dot{\alpha}_{i,m-1} + H_{i,m}(t)
$$

where

$$
\dot{\alpha}_{i,m-1} = \sum_{\vartheta=1}^{m-1}\frac{\partial\alpha_{i,m-1}}{\partial x_{i,\vartheta}}[x_{i,\vartheta+1} + f_{i,\vartheta}(x_i) + H_{i,\vartheta}(t)] + \sum_{\vartheta=0}^{m-1}\frac{\partial\alpha_{i,m-1}}{\partial y_0^{(\vartheta)}}y_0^{(\vartheta+1)}
$$

$$
\begin{aligned}
&+\sum_{\vartheta=1}^{m-1}\sum_{j\in N_i}\frac{\partial\alpha_{i,m-1}}{\partial x_{j,\vartheta}}[x_{j,\vartheta+1}+f_{j,\vartheta}(x_j)+H_{j,\vartheta}(t)]+\frac{\partial\alpha_{i,m-1}}{\partial\theta_i}\dot{\hat{\theta}}_i\\
&+\sum_{\vartheta=1}^{m-1}\frac{\partial\alpha_{i,m-1}}{\partial\hat{z}_{i,\vartheta}}\lambda_{i,\vartheta}(\tilde{z}_{i,\vartheta}+\tilde{M}_{i,\vartheta}^T S_{i,\vartheta}(x_i)+\upsilon_{i,\vartheta}(x_i))\\
&+\sum_{\vartheta=1}^{m-1}\sum_{j\in N_i}\frac{\partial\alpha_{i,m-1}}{\partial\hat{z}_{j,\vartheta}}\lambda_{j,\vartheta}(\tilde{z}_{j,\vartheta}+\tilde{M}_{j,\vartheta}^T S_{j,\vartheta}(x_j)+\upsilon_{j,\vartheta}(x_j)).
\end{aligned}
$$

Define

$$
V_{i,m}=V_{i,m-1}+\frac{1}{2}e_{i,m}^2+\frac{1}{2}\tilde{z}_{i,m}^2+\frac{1}{2\eta_{i,m}}\tilde{M}_{i,m}^T\tilde{M}_{i,m}
$$

where $\eta_{i,m}>0$. The $\dot{V}_{i,m}$ is derived as

$$
\begin{aligned}
\dot{V}_{i,m}&=e_{i,m}[x_{i,m+1}+f_{i,m}(x_{i,m})+H_{i,m}(t)-\dot{\alpha}_{i,m-1}]\\
&\quad+\tilde{z}_{i,m}(-\lambda_{i,m}\tilde{z}_{i,m}+\dot{H}_{i,m}-\lambda_{i,m}\tilde{M}_{i,m}^T S_{i,m}(x_i)\\
&\quad-\lambda_{i,m}\upsilon_{i,m}(x_i))-\frac{1}{\eta_{i,m}}\tilde{M}_{i,m}^T\dot{\hat{M}}_{i,m}+\dot{V}_{i,m-1}\\
&=\dot{V}_{i,m-1}+e_{i,m}(e_{i,m+1}+\alpha_{i,m}+\bar{f}_{i,m}(Z_{i,m})+\tilde{z}_{i,m})-\hat{m}_{i,m}e_{i,m}^2\\
&\quad-\hat{c}_{i,m}e_{i,m}^2-e_{i,m}\left[\sum_{\vartheta=1}^{m-1}\sum_{j\in N_i}(\frac{\partial\alpha_{i,m-1}}{\partial x_{j,\vartheta}}+\frac{\partial\alpha_{i,m-1}}{\partial\hat{z}_{j,\vartheta}}\lambda_{j,\vartheta})\tilde{z}_{j,\vartheta}\right.\\
&\quad\left.+\sum_{\vartheta=1}^{m-1}(\frac{\partial\alpha_{i,m-1}}{\partial x_{i,\vartheta}}+\frac{\partial\alpha_{i,m-1}}{\partial\hat{z}_{i,\vartheta}}\lambda_{i,\vartheta})\tilde{z}_{i,\vartheta}\right]-\lambda_{i,m}\tilde{z}_{i,m}\upsilon_{i,m}(x_i)\\
&\quad-\frac{1}{\eta_{i,m}}\tilde{M}_{i,m}^T(\dot{\hat{M}}_{i,m}+\eta_{i,m}\lambda_{i,m}\tilde{z}_{i,m}S_{i,m}(x_i))-\frac{e_{i,m}^2}{2}\\
&\quad+e_{i,m}[\varphi_{i,m}(Z_{i,m})-\frac{\partial\alpha_{i,m-1}}{\partial\theta_i}\dot{\hat{\theta}}_i]+\tilde{z}_{i,m}(-\lambda_{i,m}\tilde{z}_{i,m}+\dot{H}_{i,m})
\end{aligned}
$$

with

$$
\begin{aligned}
\bar{f}_{i,m}(Z_{i,m})&=f_{i,m}(x_i)-\sum_{\vartheta=1}^{m-1}\frac{\partial\alpha_{i,m-1}}{\partial x_{i,\vartheta}}\left[x_{i,\vartheta+1}+f_{i,\vartheta}(x_i)\right]-\sum_{\vartheta=1}^{m-1}\frac{\partial\alpha_{i,m-1}}{\partial y_0^{(\vartheta)}}y_0^{(\vartheta+1)}\\
&\quad-\sum_{\vartheta=1}^{m-1}\sum_{j\in N_i}\frac{\partial\alpha_{i,m-1}}{\partial x_{j,\vartheta}}\left[x_{i,\vartheta+1}+f_{i,\vartheta}(x_j)\right]-\sum_{\vartheta=1}^{m-1}\sum_{j\in N_i}\frac{\partial\alpha_{i,m-1}}{\partial x_{j,\vartheta}}\hat{z}_{j,\vartheta}\\
&\quad-\sum_{\vartheta=1}^{m-1}\frac{\partial\alpha_{i,m-1}}{\partial x_{i,\vartheta}}\hat{z}_{i,\vartheta}-\sum_{\vartheta=1}^{m-1}\frac{\partial\alpha_{i,m-1}}{\partial\hat{z}_{i,\vartheta}}\lambda_{i,\vartheta}(\tilde{M}_{i,\vartheta}^T S_{i,\vartheta}(x_i)+\upsilon_{i,\vartheta}(x_i))
\end{aligned}
$$

$$-\varphi_{i,m}(Z_{i,m}) - \sum_{\vartheta=1}^{m-1}\sum_{j\in N_i}\frac{\partial\alpha_{i,m-1}}{\partial\hat{z}_{j,\vartheta}}\lambda_{j,\vartheta}(\tilde{M}_{j,\vartheta}^T S_{j,\vartheta}(x_j) + \upsilon_{j,\vartheta}(x_j))$$
$$+\frac{e_{i,m}}{2} + (\tilde{d}_i + \tilde{a}_{i,0})e_{i,m-1} + \hat{c}_{i,m}e_{i,m} + \hat{m}_{i,m}e_{i,m} + \hat{z}_{i,m}$$

where for $m = 2$, $\tilde{d}_i + \tilde{a}_{i,0} = d_i + a_{i,0}$, and for $m > 2$, $\tilde{d}_i + \tilde{a}_{i,0} = 1$. Define $c_i > 0$ and $m_i > 0$, let

$$\hat{c}_{i,m} = \sum_{\vartheta=1}^{m-1}\left(\frac{1}{4c_i}\right)\left[\left(\frac{\partial\alpha_{i,m-1}}{\partial x_{i,\vartheta}}\right) + \left(\frac{\partial\alpha_{i,m-1}}{\partial\hat{z}_{j,m}}\right)\lambda_{i,\vartheta}\right]^2$$
$$\hat{m}_{i,m} = \sum_{\vartheta=1}^{m-1}\sum_{j\in N_i}\left(\frac{1}{4m_i}\right)\left[\left(\frac{\partial\alpha_{i,m-1}}{\partial x_{j,\vartheta}}\right) + \left(\frac{\partial\alpha_{i,m-1}}{\partial\hat{z}_{j,\vartheta}}\right)\lambda_{j,\vartheta}\right]^2.$$

According to Young's inequality, it has

$$-e_{i,m}\sum_{\vartheta=1}^{m-1}\left(\frac{\partial\alpha_{i,m-1}}{\partial x_{i,\vartheta}} + \frac{\partial\alpha_{i,m-1}}{\partial\hat{z}_{i,\vartheta}}\lambda_{i,\vartheta}\right)\tilde{z}_{i,\vartheta} \le \hat{c}_{i,m}e_{i,m}^2 + \sum_{\vartheta=1}^{m-1}c_i\tilde{z}_{i,\vartheta}^2$$
$$-e_{i,m}\sum_{\vartheta=1}^{m-1}\sum_{j\in N_i}\left(\frac{\partial\alpha_{i,m-1}}{\partial x_{j,\vartheta}} + \frac{\partial\alpha_{i,m-1}}{\partial\hat{z}_{j,\vartheta}}\lambda_{j,\vartheta}\right)\tilde{z}_{j,\vartheta} \le \hat{m}_{i,m}e_{i,m}^2 + \sum_{\vartheta=1}^{m-1}\sum_{j\in N_i}m_i\tilde{z}_{j,\vartheta}^2$$
$$e_{i,m}\tilde{z}_{i,m} \le \frac{e_{i,m}^2}{2} + \frac{\tilde{z}_{i,m}^2}{2}$$
$$-\lambda_{i,m}\tilde{z}_{i,m}\upsilon_{i,m}(x_i) \le \frac{1}{2}\tilde{z}_{i,m}^2 + \frac{\lambda_{i,m}^2\upsilon_{i,m,0}^2}{2}.$$

The function $\varphi_{i,m}(Z_{i,m})(2 \le m \le n-1)$ is used to estimate $(\frac{\partial\alpha_{i,m-1}}{\partial\hat{\theta}_i})\dot{\hat{\theta}}_i$, where

$$\varphi_{i,m}(Z_{i,m}) = -\rho_i\hat{\theta}_i\frac{\partial\alpha_{i,m-1}}{\partial\hat{\theta}_i} - \sum_{\vartheta=2}^{m}\frac{e_{i,m}\gamma_i}{2p_{i,m}^2}\left|e_{i,m}\frac{\partial\alpha_{i,\vartheta-1}}{\partial\hat{\theta}_i}\right|$$
$$+\sum_{\vartheta=1}^{m-1}\frac{\partial\alpha_{i,m-1}}{\partial\hat{\theta}_i}\frac{\gamma_i}{2p_{i,\vartheta}^2}e_{i,\vartheta}^2 S_{i,\vartheta}^T S_{i,\vartheta}.$$

The RBF NNs are used to compensate $\bar{f}_{i,m}$ with

$$Z_{i,m} = \left[x_i^T, x_j^T, \hat{y}_0, \hat{\theta}_i, \hat{z}_{i,m}^T, \hat{z}_{j,m}^T\right]^T \quad (j \in N_i).$$

Hence, one has

$$\bar{f}_{i,m} = W_{i,m}^{*T}S_{i,m}(Z_{i,m}) + F_{i,m}(Z_{i,m})$$

where $0 < \left|F_{i,m}(Z_{i,m})\right| \leq \varepsilon_{i,m}$. According to Young's inequality, we can get

$$e_{i,m}\bar{f}_{i,m}(Z_{i,m}) \leq \frac{\theta_{i,m}}{2p_{i,m}^2}e_{i,m}^2 S_{i,m}^T(\bar{Z}_{i,m})S_{i,m}(\bar{Z}_{i,m}) + \frac{p_{i,m}^2}{2} + \frac{e_{i,m}^2}{2} + \frac{\varepsilon_{i,m}^2}{2}$$

where $p_{i,m} > 0$ and $\bar{Z}_{i,m} = [x_{i,m}^T, x_{j,m}^T, \hat{y}_0^T, \hat{\theta}_i, \hat{z}_{i,m}^T, \hat{z}_{j,m}^T]^T$. Design

$$\alpha_{i,m} = -k_{i,m}e_{i,m} - \frac{e_{i,m}}{2} - \frac{\hat{\theta}_i}{2p_{i,m}^2}e_{i,m}S_{i,m}^T(\bar{Z}_{i,m})S_{i,m}(\bar{Z}_{i,m})$$

$$\dot{\hat{M}}_{i,m} = -\eta_{i,m}\lambda_{i,m}\hat{z}_{i,m}S_{i,m}(x_i) - l_{i,m}\hat{M}_{i,m}$$

where $k_{i,m} > 0$, $p_{i,m} > 0$ and $l_{i,m} > 0$. Similarly, one obtains

$$\begin{aligned}
\dot{V}_{i,m} \leq\ & e_{i,m}e_{i,m+1} - \sum_{\vartheta=1}^{m} k_{i,\vartheta}e_{i,\vartheta}^2 + \sum_{\vartheta=1}^{m}\left(\frac{p_{i,\vartheta}^2}{2} + \frac{\varepsilon_{i,\vartheta}^2}{2}\right) + \sum_{\vartheta=1}^{m-1} e_{i,\vartheta}(m-\vartheta)c_i\tilde{z}_{i,\vartheta}^2 \\
& + \frac{\tilde{\theta}_i}{\gamma_i}\Big(\sum_{\vartheta=1}^{m}\frac{\gamma_i}{2p_{i,\vartheta}^2}e_{i,\vartheta}^2 S_{i,\vartheta}^T(\bar{Z}_{i,\vartheta})S_{i,\vartheta}(\bar{Z}_{i,\vartheta}) - \dot{\hat{\theta}}_i\Big) + \sum_{\vartheta=1}^{m-1}\sum_{j\in N_i}(m-\vartheta)m_i\tilde{z}_{j,\vartheta}^2 \\
& + \frac{(d_i + a_{i,0})}{2}\tilde{z}_{i,1}^2 + \frac{1}{2}\sum_{j\in N_i} a_{i,j}\tilde{z}_{j,1}^2 + \sum_{\vartheta=2}^{m}\frac{\tilde{z}_{i,\vartheta}^2}{2} + \sum_{\vartheta=1}^{m}\frac{\lambda_{i,\vartheta}^2 \upsilon_{i,\vartheta,0}^2}{2} \\
& + \sum_{\vartheta=1}^{m}\tilde{z}_{i,\vartheta}(-\lambda_{i,\vartheta}\tilde{z}_{i,\vartheta} + \frac{1}{2}\tilde{z}_{i,\vartheta} + \dot{H}_{i,\vartheta}) + \sum_{\vartheta=1}^{m}\frac{l_{i,\vartheta}}{\eta_{i,\vartheta}}\tilde{M}_{i,\vartheta}^T\hat{M}_{i,\vartheta} \\
& + \sum_{\vartheta=1}^{m} e_{i,\vartheta}[\varphi_{i,m}(Z_{i,m}) - \frac{\partial\alpha_{i,m-1}}{\partial\theta_i}\dot{\hat{\theta}}_i].
\end{aligned}$$

Step n: The dynamics of $e_{i,n}$ is shown as

$$\dot{e}_{i,n} = \sum_{p=1}^{M} b_{i,p}(k_{i,p}u_{i,p} + \bar{u}_{i,p}) + f_{i,n}(x_i) + H_{i,n}(t) - \dot{\alpha}_{i,n-1}$$

where

$$\dot{\alpha}_{i,n-1} = \sum_{\vartheta=1}^{n-1}\sum_{j\in N_i}\frac{\partial\alpha_{i,n-1}}{\partial\hat{z}_{j,\vartheta}}\lambda_{i,\vartheta}(\tilde{z}_{i,\vartheta} + M_{j,\vartheta}^{*T}S_{j,\vartheta}\left(x_j\right) + \upsilon_{j,\vartheta}\left(x_j\right))$$

$$+\sum_{\vartheta=1}^{n-1}\frac{\partial\alpha_{i,n-1}}{\partial\hat{z}_{i,\vartheta}}\lambda_{i,\vartheta}(\tilde{z}_{i,\vartheta}+M_{i,\vartheta}^{*T}S_{i,\vartheta}(x_i)+\upsilon_{i,\vartheta}(x_i))$$
$$+\sum_{\vartheta=1}^{n-1}\frac{\partial\alpha_{i,n-1}}{\partial x_{i,\vartheta}}[x_{i,\vartheta+1}+f_{i,\vartheta}(x_i)+H_{i,\vartheta}(t)]+\sum_{\vartheta=0}^{n-1}\frac{\partial\alpha_{i,n-1}}{\partial y_0^{(\vartheta)}}y_0^{(\vartheta+1)}$$
$$+\sum_{\vartheta=1}^{n-1}\sum_{j\in N_i}\frac{\partial\alpha_{i,n-1}}{\partial x_{i,\vartheta}}[x_{j,\vartheta+1}+f_{j,\vartheta}(x_i)+H_{j,\vartheta}(t)]+\frac{\partial\alpha_{i,n-1}}{\partial\hat{\theta}_i}\dot{\hat{\theta}}_i.$$

Then, the triggering event is defined for the pth actuator as

$$u_{i,p}(t)=\omega_{i,p}(t_k^p),\forall t\in[t_k^p,t_{k+1}^p)\tag{7.3}$$
$$t_{k+1}^p=\inf\{t>t_k^p\,||\Lambda_{i,p}|\geq\Delta_{i,p}|u_{i,p}|+\epsilon_{i,p}\}\tag{7.4}$$

where the mechanism error is $\Lambda_{i,p}=\omega_{i,p}(t)-u_{i,p}(t)$, $0<\Delta_{i,p}<1$ and $\epsilon_{i,p}>0$ with $p=1,\ldots,M$.

Remark 7.2 For $\forall t\in\left[t_k^p,t_{k+1}^p\right)$, the p-th actuator's input signal holds $\omega_{i,p}(t_k^p)$. If (7.4) is triggered, the time instant will be marked as t_{k+1}^p.

Furthermore, $u_{i,p}$ can be presented as

$$u_{i,p}=\frac{\omega_{i,p}-\tau_{i,2}^p\epsilon_{i,p}}{1+\tau_{i,1}^k\Delta_{i,p}}$$

where $\left|\tau_{i,1}^k(t)\right|\leq1$ and $\left|\tau_{i,2}^p(t)\right|\leq1$. Let $\Pi_i=\inf_{t\geq0}\sum_{p=1}^{m}b_{i,p}k_{i,p}$, $\Upsilon_i=(\frac{1}{\Pi_i})$, $P_i=\Upsilon_i\sup_{t\geq0}\|\varpi_i\|$ with $\varpi_i=\left[b_{i,1}\bar{u}_{i,1}-\frac{b_{i,p}\tau_{i,2}^p\epsilon_{i,1}}{1+\tau_{i,1}^{\kappa}\Delta_{i,1}},\ldots,b_{i,p}\bar{u}_{i,p}-\frac{b_{i,p}\tau_{i,2}^p\epsilon_{i,p}}{1+\tau_{i,1}^{\kappa}\Delta_{i,p}}\right]^T$. Moreover, define $\bar{I}=[1,\ldots,1]^T$. The Lyapunov function $V_{i,n}$ is chosen as

$$V_{i,n}=V_{i,n-1}+\frac{1}{2}e_{i,n}^2+\frac{\tilde{z}_{i,n}^2}{2}+\frac{\Pi_i}{2r_{i,1}}\tilde{P}_i^2+\frac{\Pi_i}{2h_{i,1}}\tilde{\Upsilon}_i^2+\frac{1}{2\eta_{i,n}}\tilde{M}_{i,n}^T\tilde{M}_{i,n}$$

where $\tilde{\Upsilon}_i=\Upsilon_i-\hat{\Upsilon}_i$, $\tilde{P}_i=P_i-\hat{P}_i$ with $\hat{\Upsilon}_i$ and $\hat{P}_i$ being the estimations of Υ_i and P_i, respectively. In this step, the adaptive controller is given as

$$\omega_{i,p}=-(1+\Delta_{i,p})\left(\hat{\Upsilon}_i\alpha_{i,n}\tanh\left(\frac{\hat{\Upsilon}_i\alpha_{i,n}}{\nu_i}\right)+\hat{P}_i\tanh\left(\frac{e_{i,n}\hat{P}_i}{\nu_i}\right)\right)\tag{7.5}$$

with

$$\alpha_{i,n}=-k_{i,n}e_{i,n}-\frac{e_{i,n}}{2}-\frac{\hat{\theta}_i}{2p_{i,n}^2}e_{i,n}S_{i,n}^T\left(\bar{Z}_{i,n}\right)S_{i,n}(\bar{Z}_{i,n})$$

where $k_{i,n}$ and $p_{i,n}^2$ are positive constants. The derivative of $V_{i,n}$ is given by

$$\dot{V}_{i,n} \leq \dot{V}_{i,n-1} + e_{i,n}\sum_{p=1}^{M} b_{i,p}\frac{\omega_{i,p}}{1+\tau_{i,1}^{p}\Delta_{i,p}} + e_{i,n}(f_{i,n}(x_i) + H_{i,n}(t) - \dot{\alpha}_{i,n-1})$$
$$+\tilde{z}_{i,n}(-\lambda_{i,n}(\tilde{M}_{i,n}^T S_{i,n}(x_i) + \upsilon_{i,n}(x_i)) + \dot{H}_{i,n} - \lambda_{i,n}\tilde{z}_{i,n}) - \frac{\Pi_i}{h_{i,1}}\tilde{\Upsilon}_i\dot{\hat{\Upsilon}}_i$$
$$+e_{i,n}\varpi_i^T\bar{I} - \frac{\Pi_i}{r_{i,1}}\tilde{P}_i\dot{\hat{P}}_i - \frac{1}{\eta_{i,n}}\tilde{M}_{i,n}^T\dot{\hat{M}}_{i,n}.$$

Then, we have

$$\dot{V}_{i,n} \leq \dot{V}_{i,n-1} + e_{i,n}\sum_{p=1}^{M} b_{i,p}\frac{\omega_{i,p}}{1+\tau_{i,1}^{p}\Delta_{i,p}} + e_{i,n}\left[\varphi_{i,n}(Z_{i,n}) - \frac{\partial\alpha_{i,n-1}}{\partial\theta_i}\dot{\hat{\theta}}_i\right]$$
$$-\hat{c}_{i,n}e_{i,n}^2 - \hat{m}_{i,n}e_{i,n}^2 - e_{i,n}\left[\sum_{\vartheta=1}^{n-1}\left(\frac{\partial\alpha_{i,n-1}}{\partial x_{i,\vartheta}} + \frac{\partial\alpha_{i,n-1}}{\partial\hat{z}_{i,\vartheta}}\lambda_{i,\vartheta}\right)\tilde{z}_{i,\vartheta}\right.$$
$$\left.+\sum_{\vartheta=1}^{n-1}\sum_{j\in N_i}\left(\frac{\partial\alpha_{i,n-1}}{\partial x_{j,\vartheta}} + \frac{\partial\alpha_{i,n-1}}{\partial\hat{z}_{j,\vartheta}}\lambda_{j,\vartheta}\right)\tilde{z}_{j,\vartheta}\right] - \lambda_{i,n}\tilde{z}_{i,n}\upsilon_{i,n}(x_i)$$
$$-\frac{1}{\eta_{i,n}}\tilde{M}_{i,n}^T(\dot{\hat{M}}_{i,n} + \eta_{i,n}\lambda_{i,n}\tilde{z}_{i,n}S_{i,n}(x_i)) + \tilde{z}_{i,n}(-\lambda_{i,n}\tilde{z}_{i,n} + \dot{H}_{i,n})$$
$$-\frac{e_{i,n}}{2} - \frac{\Pi_i}{h_{i,1}}\tilde{\Upsilon}_i\dot{\hat{\Upsilon}}_i + e_{i,n}\varpi_i^T\bar{I} + e_{i,n}\bar{f}_{i,n}(Z_{i,n}) \tag{7.6}$$

with

$$\bar{f}_{i,n}(Z_{i,n}) = f_{i,n}(x_i) - \sum_{\vartheta=1}^{n-1}\frac{\partial\alpha_{i,n-1}}{\partial x_{i,\vartheta}}\left[x_{i,\vartheta+1} + f_{i,\vartheta}(x_i)\right] - \sum_{\vartheta=1}^{n-1}\frac{\partial\alpha_{i,n-1}}{\partial x_{i,\vartheta}}\hat{z}_{i,\vartheta}$$
$$-\sum_{\vartheta=1}^{n-1}\sum_{j\in N_i}\frac{\partial\alpha_{i,n-1}}{\partial x_{j,\vartheta}}\left[x_{i,\vartheta+1} + f_{i,\vartheta}(x_j)\right] - \sum_{\vartheta=1}^{n-1}\frac{\partial\alpha_{i,n-1}}{\partial y_0^{(\vartheta)}}y_0^{(\vartheta+1)}$$
$$-\varphi_{i,n}(Z_{i,n}) - \sum_{\vartheta=1}^{n-1}\frac{\partial\alpha_{i,n-1}}{\partial\hat{z}_{i,\vartheta}}\lambda_{i,\vartheta}(\tilde{M}_{i,\vartheta}^T S_{i,\vartheta}(x_i) + \upsilon_{i,\vartheta}(x_i))$$
$$-\sum_{\vartheta=1}^{n-1}\sum_{j\in N_i}\frac{\partial\alpha_{i,n-1}}{\partial\hat{z}_{j,\vartheta}}\lambda_{j,\vartheta}(\tilde{M}_{j,\vartheta}^T S_{j,\vartheta}(x_j) + \upsilon_{j,\vartheta}(x_j)) + e_{i,n-1}$$
$$-\sum_{\vartheta=1}^{n-1}\sum_{j\in N_i}\frac{\partial\alpha_{i,n-1}}{\partial x_{j,\vartheta}}\hat{z}_{j,\vartheta} + \frac{e_{i,n}}{2} + \hat{c}_{i,n}e_{i,n} + \hat{m}_{i,n}e_{i,n} + \hat{z}_{i,n}.$$

Further, one has

$$e_{i,n}\varpi_i^T\bar{I} \le \Pi_i P_i \left|e_{i,n}\right| = \Pi_i \tilde{P}_i \left|e_{i,n}\right| + \Pi_i \hat{P}_i \left|e_{i,n}\right|.$$

Then, according to $0 < |g| - F\tanh(g/\nu_i) \le 0.2785\nu_i$, one has

$$\Pi_i \hat{P}_i \left|e_{i,n}\right| \le \Pi_i\left(e_{i,n}\hat{P}_i \tanh\left(\frac{e_{i,n}\hat{P}_i}{\nu_i}\right) + 0.2785\nu_i\right)$$

where $\nu_i > 0$ and $g \in R$. It can be obtained that

$$\begin{aligned}
&e_{i,n}\sum_{p=1}^{M} b_{i,p}\frac{\omega_{i,p}}{1+\tau_{i,1}^{p}\Delta_{i,p}}\\
&\le -\Pi_i e_{i,n}\left(\hat{\Upsilon}_i\alpha_{i,n}\tanh\left(\frac{\hat{\Upsilon}_i\alpha_{i,n}}{\nu_i}\right) + \hat{P}_i\tanh\left(\frac{e_{i,n}\hat{P}_i}{\nu_i}\right)\right)\\
&\le -\Pi_i\hat{\Upsilon}_i\left|e_{i,n}\alpha_{i,n}\right| + 0.2785\nu_i\Pi_i - \Pi_i e_{i,n}\hat{P}_i\tanh\left(\frac{e_{i,n}\hat{P}_i}{\nu_i}\right).
\end{aligned}$$

Furthermore, we have

$$-\Pi_i\hat{\Upsilon}_i\left|e_{i,n}\alpha_{i,n}\right| = \Pi_i\tilde{\Upsilon}_i\left|e_{i,n}\alpha_{i,n}\right| - \left|e_{i,n}\alpha_{i,n}\right|.$$

Let

$$\begin{aligned}
\hat{c}_{i,n} &= \sum_{\vartheta=1}^{n-1}\left(\frac{1}{4c_i}\right)\left[\left(\frac{\partial\alpha_{i,n-1}}{\partial x_{i,\vartheta}}\right) + \left(\frac{\partial\alpha_{i,n-1}}{\partial\hat{z}_{j,n}}\right)\lambda_{i,\vartheta})\right]^2\\
\hat{m}_{i,n} &= \sum_{\vartheta=1}^{n-1}\sum_{j\in N_i}\left(\frac{1}{4m_i}\right)\left[\left(\frac{\partial\alpha_{i,n-1}}{\partial x_{j,\vartheta}}\right) + \left(\frac{\partial\alpha_{i,n-1}}{\partial\hat{z}_{j,\vartheta}}\right)\lambda_{j,\vartheta}\right]^2.
\end{aligned}$$

According to Young's inequality, it has

$$\begin{aligned}
-e_{i,n}\sum_{\vartheta=1}^{n-1}\left(\frac{\partial\alpha_{i,n-1}}{\partial x_{i,\vartheta}} + \frac{\partial\alpha_{i,n-1}}{\partial\hat{z}_{i,\vartheta}}\lambda_{i,\vartheta}\right)\tilde{z}_{i,\vartheta} &\le \hat{c}_{i,n}e_{i,n}^2 + \sum_{\vartheta=1}^{n-1}c_i\tilde{z}_{i,\vartheta}^2\\
-e_{i,n}\sum_{\vartheta=1}^{n-1}\sum_{j\in N_i}\left(\frac{\partial\alpha_{i,n-1}}{\partial x_{j,\vartheta}} + \frac{\partial\alpha_{i,n-1}}{\partial\hat{z}_{j,\vartheta}}\lambda_{j,\vartheta}\right)\tilde{z}_{j,\vartheta} &\le \hat{m}_{i,n}e_{i,n}^2 + \sum_{\vartheta=1}^{n-1}\sum_{j\in N_i}m_i\tilde{z}_{j,\vartheta}^2\\
e_{i,n}\tilde{z}_{i,n} &\le \frac{e_{i,n}^2}{2} + \frac{\tilde{z}_{i,n}^2}{2}\\
-\lambda_{i,n}\tilde{z}_{i,n}\upsilon_{i,n}(x_i) &\le \frac{1}{2}\tilde{z}_{i,n}^2 + \frac{\lambda_{i,n}^2\upsilon_{i,n,0}^2}{2}.
\end{aligned}$$

The function $\varphi_{i,n}(Z_{i,n})$ is used to compensate $(\frac{\partial \alpha_{i,n-1}}{\partial \hat{\theta}_i})\dot{\hat{\theta}}_i$, where

$$\varphi_{i,n}(Z_{i,n}) = -\rho_i \hat{\theta}_i \frac{\partial \alpha_{i,n-1}}{\partial \hat{\theta}_i} - \sum_{\vartheta=2}^{n} \frac{e_{i,n}\gamma_i}{2p_{i,n}^2} \left| e_{i,n} \frac{\partial \alpha_{i,\vartheta-1}}{\partial \hat{\theta}_i} \right| + \sum_{\vartheta=1}^{n-1} \frac{\partial \alpha_{i,n-1}}{\partial \hat{\theta}_i} \frac{\gamma_i}{2p_{i,\vartheta}^2} e_{i,\vartheta}^2 S_{i,\vartheta}^T S_{i,\vartheta}.$$

The RBF NNs are employed to approximate $\bar{f}_{i,n}$:

$$\bar{f}_{i,n} = W_{i,n}^{*T} S_{i,n}(Z_{i,n}) + \delta_{i,n}(Z_{i,n})$$

where $0 < \left|\delta_{i,n}(Z_{i,n})\right| \leq \varepsilon_{i,n}$, $Z_{i,n} = \left[x_i^T, x_j^T, \hat{y}_0, \hat{\theta}_i, \hat{z}_{i,n}^T, \hat{z}_{j,n}^T\right]^T$ $(j \in N_i)$. Similarly, it yields,

$$e_{i,n}\bar{f}_{i,n}(Z_{i,n}) \leq \frac{\theta_{i,n}}{2p_{i,n}^2} e_{i,n}^2 S_{i,n}^T(\bar{Z}_{i,n}) S_{i,n}(\bar{Z}_{i,n}) + \frac{p_{i,n}^2}{2} + \frac{e_{i,n}^2}{2} + \frac{\varepsilon_{i,n}^2}{2}$$

where $p_{i,n} > 0$ and $Z_{i,n} = \bar{Z}_{i,n}$. The adaptive laws are given as

$$\begin{aligned}
\dot{\hat{\theta}}_i &= \sum_{\vartheta=1}^{n} \frac{\gamma_i}{2p_{i,\vartheta}^2} e_{i,\vartheta} S_{i,\vartheta}^T \left(Z_{i,\vartheta}\right) S_{i,\vartheta}(Z_{i,\vartheta}) - \rho_i \hat{\theta}_i \\
\dot{\hat{M}}_{i,n} &= -\eta_{i,n}\lambda_{i,n}\hat{z}_{i,n} S_{i,n}(x_i) - l_{i,n}\hat{M}_{i,n} \\
\dot{\hat{\Upsilon}}_i &= h_{i,1}\left|e_{i,n}\alpha_{i,n}\right| - h_{i,2}\hat{\Upsilon}_i \\
\dot{\hat{P}}_i &= \gamma_{i,1}\left|e_{i,n}\right| - \gamma_{i,2}\hat{P}_i
\end{aligned} \tag{7.7}$$

where $\eta_{i,n}$, $l_{i,n}$, ρ_i, $h_{i,1}$, $h_{i,2}$, $g_{i,1}$ and $g_{i,2}$ are positive parameters. The following inequality holds

$$\begin{aligned}
\dot{V}_{i,n} \leq &-\sum_{\vartheta=1}^{n} k_{i,\vartheta} e_{i,\vartheta}^2 + 0.557\nu_i \Pi_i + \frac{h_{i,2}\Pi_i}{h_{i,1}} \tilde{\Upsilon}_i \hat{\Upsilon}_i + \frac{(d_i + a_{i,0}) + 1}{2} \tilde{z}_{i,1}^2 \\
&+ \frac{\gamma_{i,2}\Pi_i}{\gamma_{i,1}} \tilde{P}_i \hat{P}_i + \frac{\rho_i}{\gamma_i} \tilde{\theta}_i \hat{\theta}_i + \sum_{\vartheta=1}^{n-1} (n-\vartheta) c_i \tilde{z}_{i,\vartheta}^2 + \frac{1}{2} \sum_{j \in N_i} a_{i,j} \tilde{z}_{j,1}^2 \\
&+ \sum_{\vartheta=1}^{n-1} \sum_{j \in N_i} (n-\vartheta) r_i \tilde{z}_{j,\vartheta}^2 + \sum_{\vartheta=1}^{n-1} \tilde{z}_{i,\vartheta}(-\lambda_{i,\vartheta}\tilde{z}_{i,\vartheta} + \dot{H}_{i,\vartheta}) + \sum_{\vartheta=2}^{n} \tilde{z}_{i,\vartheta}^2 \\
&+ \sum_{\vartheta=1}^{n} \left(\frac{p_{i,\vartheta}^2}{2} + \frac{\varepsilon_{i,\vartheta}^2}{2} \right) + \sum_{\vartheta=1}^{n} \frac{\lambda_{i,\vartheta}^2 \upsilon_{i,\vartheta,0}^2}{2} + \sum_{\vartheta=1}^{n} \frac{l_{i,\vartheta}}{\eta_{i,\vartheta}} \tilde{M}_{i,\vartheta}^T \hat{M}_{i,\vartheta}.
\end{aligned} \tag{7.8}$$

Theorem 7.1 *Considering the MASs with actuator faults, sensor faults, and unknown disturbances* (7.1)*, under the event-triggering rules* (7.3) *and* (7.4)*, the event-triggered controller* (7.5)*, and the adaptive laws* (7.7)*, then, it can be realized that all the signals are semi-globally uniformly ultimately bounded and the tracking errors arrive at a bounded compact set of the origin.*

Proof Design the Lyapunov function as

$$V = \sum_{i=1}^{N} V_{i,n}.$$

According to the Young's inequality, we have

$$\begin{aligned}
\tilde{\theta}_i\hat{\theta}_i &\leq -\frac{1}{2}\tilde{\theta}_i^2 + \frac{1}{2}\theta_i^2 \\
\tilde{M}_{i,\vartheta}^T\hat{M}_{i,\vartheta} &\leq -\frac{1}{2}\tilde{M}_{i,\vartheta}^T\tilde{M}_{i,\vartheta} + \frac{1}{2}M_{i,\vartheta}^{*T}M_{i,\vartheta}^{*} \\
\frac{h_{i,2}\Pi_i}{h_{i,1}}\tilde{\Upsilon}_i\hat{\Upsilon}_i &\leq \frac{h_{i,2}\Pi_i}{2h_{i,1}}\Upsilon_i^2 - \frac{h_{i,2}\Pi_i}{2h_{i,1}}\tilde{\Upsilon}_i^2 \\
\frac{\gamma_{i,2}\Pi_i}{\gamma_{i,1}}\tilde{P}_i\hat{P}_i &\leq \frac{\gamma_{i,2}\Pi_i}{2\gamma_{i,1}}P_i^2 - \frac{\gamma_{i,2}\Pi_i}{2\gamma_{i,1}}\tilde{P}_i^2.
\end{aligned}$$

The derivative of V can be described as

$$\begin{aligned}
\dot{V} \leq & \sum_{i=1}^{N}\sum_{\vartheta=1}^{n} k_{i,\vartheta}e_{i,\vartheta}^2 - \sum_{i=1}^{N}\frac{\rho_i}{2\gamma_i}\tilde{\theta}_i^2 - \sum_{i=1}^{N}\frac{\gamma_{i,2}\Pi_i}{2\gamma_{i,1}}\tilde{P}_i^2 - \sum_{i=1}^{N}\frac{h_{i,2}\Pi_i}{2h_{i,1}}\tilde{\Upsilon}_i^2 + \frac{h_{i,2}\Pi_i}{2h_{i,1}}\Upsilon_i^2 \\
& -\sum_{i=1}^{N}\Bigg\{\Bigg[\lambda_{i,1} - \sum_{j\in N_i}\frac{a_{i,j}}{2} - (n-1)c_i - \sum_{j\in N_i}(n-1)r_j - \frac{d_i + a_{i,0} + 1}{2}\Bigg]\tilde{z}_{i,1}^2 \\
& -\sum_{q=2}^{n-1}\Bigg[\lambda_{i,q} - 1 - (n-q)c_i - \sum_{j\in N_i}(n-q)r_j\Bigg]\tilde{z}_{i,q}^2 - (\lambda_{i,n}-1)\tilde{z}_{i,n}^2\Bigg\} \\
& +\sum_{i=1}^{N}\sum_{\vartheta=1}^{n}\left(\frac{p_{i,\vartheta}^2}{2} + \frac{\varepsilon_{i,\vartheta}^2}{2} + \frac{\lambda_{i,\vartheta}^2 \upsilon_{i,\vartheta,0}^2}{2}\right) + \sum_{i=1}^{N}\frac{\eta_i}{2\gamma_i}\theta_i^2 + \sum_{i=1}^{N}0.557\nu_i\Pi_i + \frac{\gamma_{i,2}\Pi_i}{2\gamma_{i,1}}P_i^2 \\
& -\sum_{i=1}^{N}\sum_{\vartheta=1}^{n}\frac{l_{i,\vartheta}}{2\eta_{i,\vartheta}}\tilde{M}_{i,\vartheta}^T\tilde{M}_{i,\vartheta} + \sum_{i=1}^{N}\sum_{\vartheta=1}^{n}\frac{l_{i,\vartheta}}{2\eta_{i,\vartheta}}M_{i,\vartheta}^{*T}M_{i,\vartheta}^{*} + \sum_{i=1}^{N}\sum_{\vartheta=1}^{n}\tilde{z}_{i,\vartheta}\dot{H}_{i,\vartheta}.
\end{aligned}$$

Define the following inequality

$$\begin{aligned}
\lambda_{i,1} &> \sum_{j\in N_i}\frac{a_{i,j}}{2} + (n-1)c_i + \sum_{j\in N_i}(n-1)r_j + \frac{d_i + a_{i,0} + 1}{2} \\
\lambda_{i,q} &> 1 + (n-q)c_i - \sum_{j\in N_i}(n-q)r_j
\end{aligned}$$

$$\lambda_{i,n} > 1$$

when $\left\|\tilde{z}_{i,\vartheta}\right\| \geq \frac{\|\dot{H}_{i,\vartheta}\|}{\lambda_{i,0}}$, and

$$\lambda_{i,0} = \min\left\{\lambda_{i,1} - \sum_{j\in N_i}\frac{a_{i,j}}{2} - (n-1)c_i - \sum_{j\in N_i}(n-1)r_j - \frac{d_i + a_{i,0} + 1}{2},\right.$$
$$\left.\lambda_{i,q} - (n-q)c_i - \sum_{j\in N_i}(n-q)r_j - 1, \lambda_{i,n} - 1\right\}$$

where $q = 2, ..., n-1$. Define

$$\mu = \min_{1\leq i\leq N}\left\{\sum_{\vartheta=1}^{n}2k_{i,\vartheta}, \eta_i, \gamma_{i,2}\Pi_i, h_{i,2}\Pi_i, 2(\lambda_{i,1} - (n-1)c_i\right.$$
$$-\sum_{j\in N_i}(n-1)r_j - \frac{d_i + a_{i,0} + 1}{2} - \sum_{j\in N_i}\frac{a_{j,i}}{2} - \lambda_{i,0}),$$
$$2(\lambda_{i,q} - (n-q)c_i - \sum_{j\in N_i}(n-q)r_j - 1 - \lambda_{i,0}),$$
$$\left.2(\lambda_{i,n} - 1 - \lambda_{i,0}), l_{i,\vartheta}\right\}$$

and

$$\Theta = \sum_{i=1}^{N}0.557\nu_i\Pi_i + \sum_{i=1}^{N}\frac{\eta_i}{2\gamma_i}\theta_i^2 + \sum_{i=1}^{N}\sum_{\vartheta=1}^{n}(\frac{p_{i,\vartheta}^2}{2} + \frac{\varepsilon_{i,\vartheta}^2}{2})$$
$$+\sum_{i=1}^{N}\sum_{\vartheta=1}^{n}\frac{l_{i,\vartheta}}{2\eta_{i,\vartheta}}M_{i,\vartheta}^{*T}M_{i,\vartheta}^{*} + \frac{h_{i,2}\Pi_i}{2h_{i,1}}\Upsilon_i^2 + \frac{\gamma_{i,2}\Pi_i}{2\gamma_{i,1}}P_i^2$$

where $q = 2, ..., n-1$. Thus, we can obtain

$$\dot{V}(t) \leq -\mu V(t) + \Theta.$$

Then, one has

$$\frac{1}{2}e_{i,1}^2 \leq V \leq e^{-\mu t}V(0) + \frac{\Theta}{\mu}(1 - e^{-\mu t}).$$

According to Lemma 7.3, we get

$$\lim_{t\to\infty}\|y - \bar{y}_0\| \leq \frac{1}{\bar{\sigma}(L+B)}\sqrt{\frac{2\Theta}{\mu}}.$$

By recalling $\Lambda_{i,p} = \omega_{i,p}(t) - u_{i,p}(t), \forall t \in \left[t_k^p, t_{k+1}^p\right)$, we can get

$$\begin{aligned}\frac{d}{dt}|\Lambda_i(t)| &= \frac{d}{dt}(\Lambda_i(t) \times \Lambda_i(t))^{\frac{1}{2}} \\ &= \text{sign}(\Lambda_i(t))\dot{\Lambda}_i(t) \le \left|\dot{\omega}_{i,p}(t)\right|.\end{aligned}$$

Since $\omega_{i,p}(t)$ is differentiable and bounded, $\left|\dot{\omega}_{i,p}(t)\right| \le \nabla_{i,p}$, where $\nabla_{i,p} > 0$. By noting that $e_{i,p}(t_p) = 0$ and $\lim\limits_{t \to t_{k+1}} e_{i,p}(t_p) = m_{i,p}$, we obtain that the lower bound of inter-execution intervals t^* must satisfy $t^* \ge \frac{m_{i,p}}{\nabla_{i,p}}$, the Zeno-behavior is successfully avoided. □

Remark 7.3 In this chapter, the actuator faults, sensor faults and unknown disturbances have been considered in MASs with event-triggered rules, and the proposed control strategy in this chapter can also save communication resources.

7.4 Simulation Results

This chapter provides the following simulation example to verify the theoretical analysis. In this section, the dynamics of MASs are given as

$$\begin{cases} \dot{x}_{i,1} = x_{i,2} - \sin(x_{i,2})x_{i,1}^2 + H_{i,1} \\ \dot{x}_{i,2} = \sum\limits_{\kappa=1}^{2} b_{i,j}k_{i,j}u_{i,j} + \sin\left(x_{i,1}\right)x_{i,2} + H_{i,2} \\ y_r = 0.5\sin\left(2.46t\right). \end{cases}$$

The following fault model is considered:

$$u_{1,1}^F = \begin{cases} u_{1,1} & \text{if } t \in [2k, 2k+1) \\ 0.5u_{1,1} & \text{if } t \in [2k+1, 2k+2) \end{cases}$$

$$u_{1,2}^F = \begin{cases} 0.3u_{1,2} & \text{if } t \in [2k, 2k+1) \\ 0 & \text{if } t \in [2k+1, 2k+2) \end{cases}$$

$$u_{2,1}^F = \begin{cases} u_{2,1} & \text{if } t \in [2k, 2k+1) \\ 0.6u_{2,1} & \text{if } t \in [2k+1, 2k+2) \end{cases}$$

$$u_{2,2}^F = \begin{cases} u_{2,2} & \text{if } t \in [0, 1) \\ 0.2 + 0.2\sin(t) & \text{if } t \in [1, \infty) \end{cases}$$

where $k = 0, 1, \ldots$ Apparently, the above matrices are given as follows,

$$\mathcal{A} = \begin{bmatrix} 0\,1\,0\,0 \\ 1\,0\,1\,1 \\ 0\,0\,0\,0 \\ 0\,0\,1\,0 \end{bmatrix}, \mathcal{L} = \begin{bmatrix} 1 & -1 & 0 & 0 \\ -1 & 3 & -1 & -1 \\ 0 & 0 & 0 & 0 \\ 0 & 0 & -1 & 1 \end{bmatrix}$$

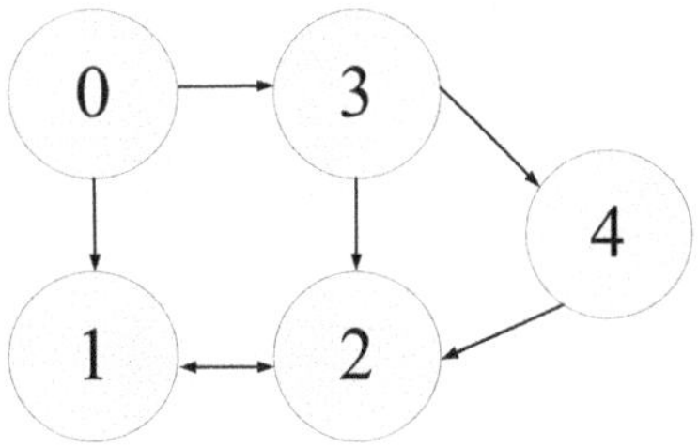

Fig. 7.1 Topology of communication graph

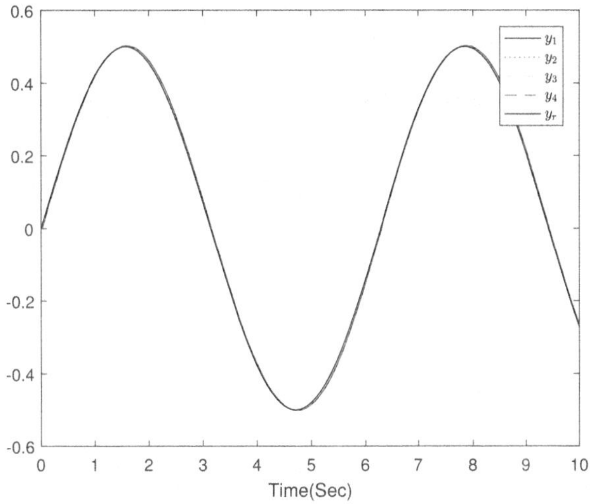

Fig. 7.2 Output trajectories of followers and the leader

From Fig. 7.1, we know $B = \mathrm{diag}(1, 0, 1, 0)$. The simulation results are shown by Figs. 7.2, 7.3, 7.4, 7.5, 7.6, and 7.7, and the correlative design parameters are chosen as $\lambda_{1,1} = 401$, $\lambda_{2,1} = 401.6$, $\lambda_{3,1} = 402.1$, $\lambda_{4,1} = 402$, $\lambda_{1,2} = 410.5$, $\lambda_{2,2} = 400.5$, $\lambda_{3,2} = 400.6$, $\lambda_{4,2} = 410.5$, $\gamma_1 = \gamma_2 = \gamma_3 = \gamma_4 = 4.96$, $k_{1,1} = 108$, $k_{2,1} = 255$, $k_{3,1} = 105$, $k_{4,1} = 109$, $k_{1,2} = k_{2,2} = k_{3,2} = k_{4,2} = 66.5$, $\rho_1 = \rho_2 = \rho_3 = \rho_4 = 0.59$, $p_{i,1} = p_{i,2} = 11.3$, $\eta_{i,1} = \eta_{i,2} = 36.4$, $l_{i,1} = 1.7$, $l_{i,2} = 1.5$, $\nu_i = 2.6$, $\Delta_i = 0.28$, $h_{i,1} = 2.71$, $h_{i,2} = 3.45$, $g_{i,1} = 2.9$, $g_{i,2} = 3.2$.

The simulation results are presented in Figs. 7.2, 7.3, 7.4, 7.5, 7.6, and 7.7. The outputs of followers can track the leader's signal in Fig. 7.2, and 7.3 shows the trajectories of tracking errors. Figure 7.4 displays the event-triggered controller. Figures 7.5 and 7.6 describe the errors of the disturbances. The trigger times and the trigger intervals of the agents are drawn in Fig. 7.7. It can be seen from the above simulation results that the event-triggered adaptive consensus of MASs with actuator faults and unknown disturbances are basically realized.

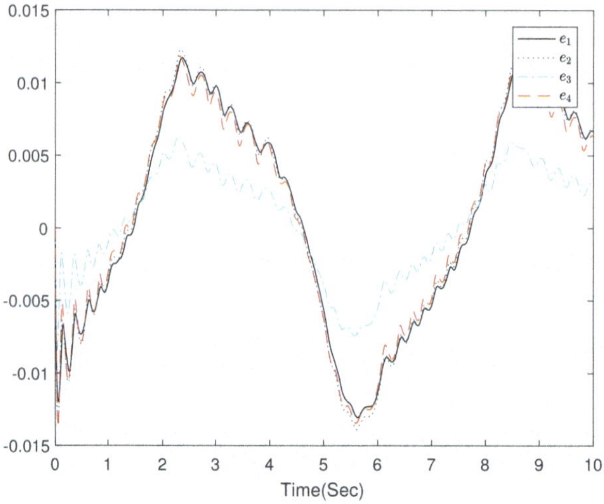

Fig. 7.3 The trajectories of tracking errors

Fig. 7.4 The trajectories of event-triggered controllers

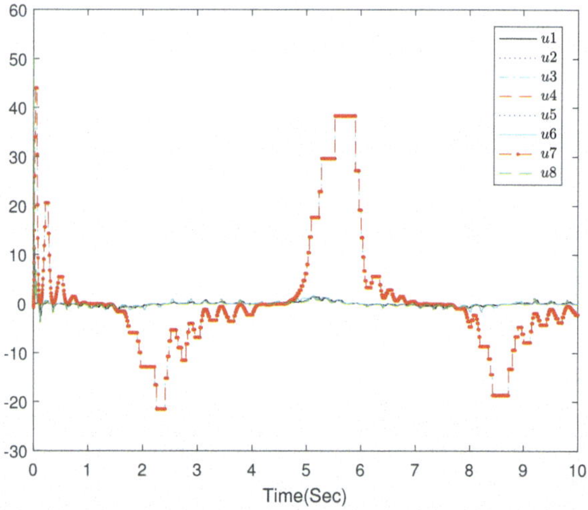

7.5 Conclusion

The consensus problem in event-triggered MASs with actuator faults and unknown disturbances has been considered in this chapter. Furthermore, the event-triggered controller has been proposed to avoid some unnecessary triggering operations. In addition, a disturbance observer has been proposed to solve the unknown disturbances. Finally, simulation results have been utilized to prove the effectiveness of

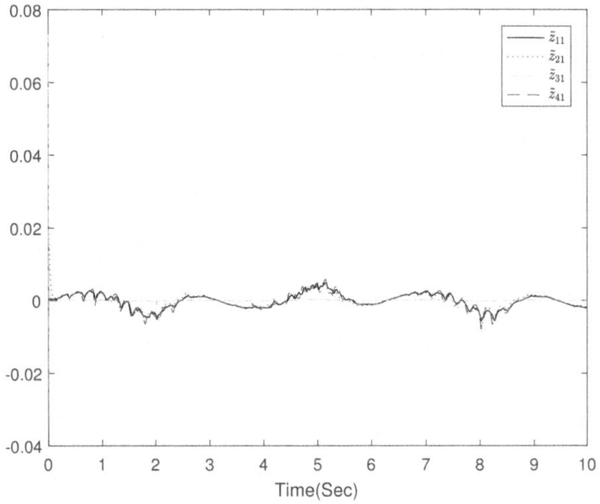

Fig. 7.5 The trajectories of errors between disturbances and disturbance observers

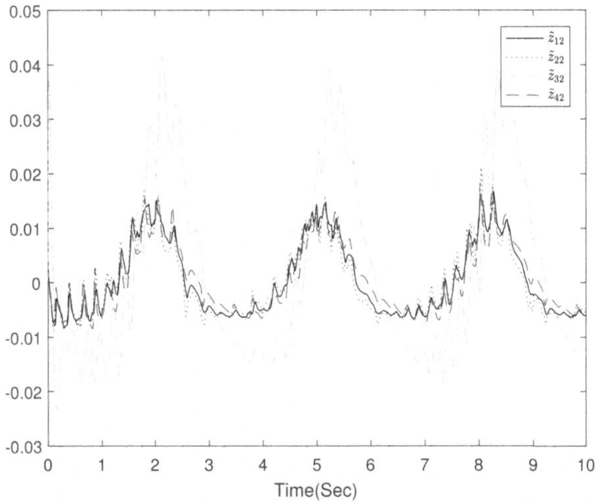

Fig. 7.6 The trajectories of errors between disturbances and disturbance observers

the proposed method. In our future research, based on the fuzzy logic system arbitrary approximation property, we will extend the results of this chapter to the fuzzy fault detection and switching topology for MASs.

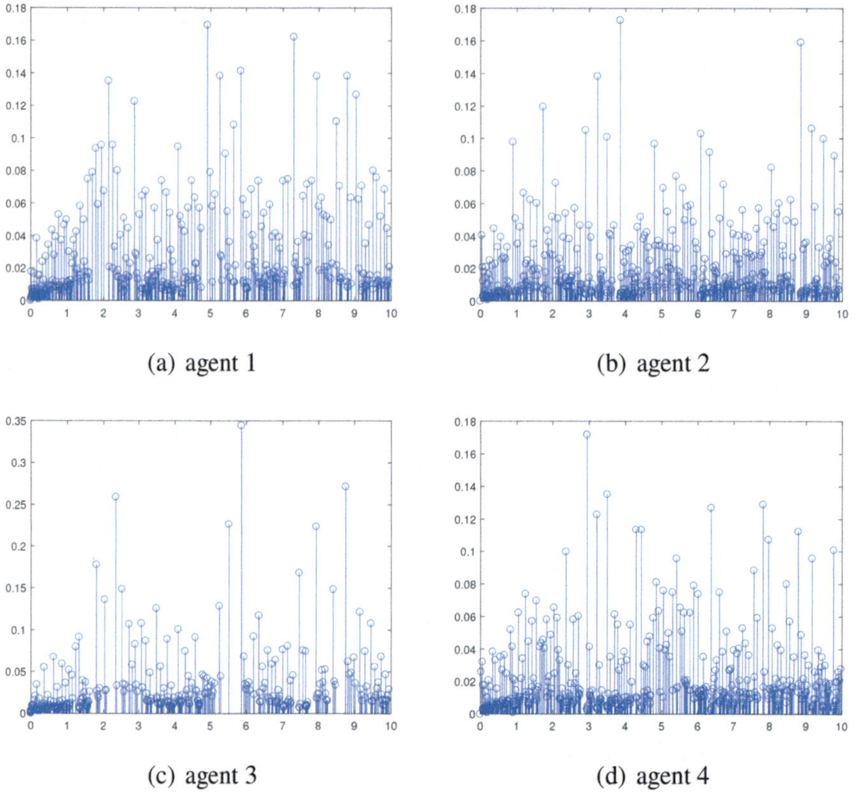

Fig. 7.7 Triggering instants and inter-event intervals

References

Bounemeur A, Chemachema M, Essounbouli N (2018) Indirect adaptive fuzzy fault-tolerant tracking control for mimo nonlinear systems with actuator and sensor failures. ISA Trans 79:45–61

Cao L, Li H, Dong G, Lu R (2021) Event-triggered control for multiagent systems with sensor faults and input saturation. IEEE Trans Syst Man Cybernet Syst 51(6):3855–3866

Chen Z, Han QL, Yan Y, Wu Z (2020) How often should one update control and estimation: review of networked triggering techniques. Sci China Inf Sci 63(5):150201

Du P, Liang H, Zhao S, Ahn CK (2021) Neural-based decentralized adaptive finite-time control for nonlinear large-scale systems with time-varying output constraints. IEEE Trans Syst Man Cybernet Syst 51(5):3136–3147

Hou Z, Cheng L, Tan M (2009) Decentralized robust adaptive control for the multiagent system consensus problem using neural networks. IEEE Trans Syst Man Cybernet Part B (Cybernetics) 39(3):636–647

Hu W, Liu L, Feng G (2016) Consensus of linear multi-agent systems by distributed event-triggered strategy. IEEE Trans Cybernet 46(1):148–157

Hu J, Wu Y, Li T, Ghosh BK (2018) Consensus control of general linear multiagent systems with antagonistic interactions and communication noises. IEEE Trans Autom Control 64(5):2122–2127

Hu W, Liu L, Feng G (2018) Event-triggered cooperative output regulation of linear multi-agent systems under jointly connected topologies. IEEE Trans Autom Control 64(3):1317–1322
Hua C, Li K, Guan X (2018) Leader-following output consensus for high-order nonlinear multiagent systems. IEEE Trans Autom Control 64(3):1156–1161
Hui M, Hongyi L, Renquan L, Huang T (2020) Adaptive event-triggered control for a class of nonlinear systems with periodic disturbances. Sci China Inf Sci 63(5):150212
Leng J, Zhang H, Yan D, Liu Q, Chen X, Zhang D (2019) Digital twin-driven manufacturing cyber-physical system for parallel controlling of smart workshop. J Ambient Intell Humanized Comput 10(3):1155–1166
Li X, Song S (2016) Stabilization of delay systems: delay-dependent impulsive control. IEEE Trans Autom Control 62(1):406–411
Li X, Wu J (2016) Stability of nonlinear differential systems with state-dependent delayed impulses. Automatica 64:63–69
Li Y, Yang G (2016) Adaptive fuzzy decentralized control for a class of large-scale nonlinear systems with actuator faults and unknown dead zones. IEEE Trans Syst Man Cybernet Syst 47(5):729–740
Li X, Ho DW, Cao J (2019) Finite-time stability and settling-time estimation of nonlinear impulsive systems. Automatica 99:361–368
Li X, Yang X, Song S (2019) Lyapunov conditions for finite-time stability of time-varying time-delay systems. Automatica 103:135–140
Liang H, Zhang L, Sun Y, Huang T (2021) Containment control of semi-Markovian multiagent systems with switching topologies. IEEE Trans Syst Man Cybernet Syst 51(6):3889–3899
Liu Y, Liu X, Jing Y, Zhang Z (2018) A novel finite-time adaptive fuzzy tracking control scheme for nonstrict feedback systems. IEEE Trans Fuzzy Syst 27(4):646–658
Liu Q, Zhang H, Leng J, Chen X (2019) Digital twin-driven rapid individualised designing of automated flow-shop manufacturing system. Int J Prod Res 57(12):3903–3919
Liu Y, Liu X, Jing Y, Chen X, Qiu J (2020) Direct adaptive preassigned finite-time control with time-delay and quantized input using neural network. IEEE Trans Neural Netw Learn Syst 31(4):1222–1231
Lu R, Xu Y, Xue A, Zheng J (2012) Networked control with state reset and quantized measurements: observer-based case. IEEE Trans Indus Electron 60(11):5206–5213
Lu R, Yu W, Lü J, Xue A (2014) Synchronization on complex networks of networks. IEEE Trans Neural Netw Learn Syst 25(11):2110–2118
Luo S, Ye D (2019) Adaptive double event-triggered control for linear multi-agent systems with actuator faults. IEEE Trans Circuits Syst I: Regular Pap 66(12):4829–4839
Qian Y, Liu L, Feng G (2018) Output consensus of heterogeneous linear multi-agent systems with adaptive event-triggered control. IEEE Trans Autom Control 64(6):2606–2613
Ren W (2010) Distributed attitude alignment in spacecraft formation flying. Int J Adapt Control Signal Process 21(2–3):95–113
Saber RO, Murray RM (2023) Consensus protocols for networks of dynamic agents. 2003 Am Control Conf 2:951–956
Sakthivel R, Parivallal A, Kaviarasan B, Lee H, Lim Y (2018) Finite-time consensus of Markov jumping multi-agent systems with time-varying actuator faults and input saturation. ISA Trans 83:89–99
Sakthivel R, Sakthivel R, Kaviarasan B, Lee H, Lim Y (2019) Finite-time leaderless consensus of uncertain multi-agent systems against time-varying actuator faults. Neurocomputing 325:159–171
Su Y, Chen B, Lin C, Wang H, Zhou S (2016) Adaptive neural control for a class of stochastic nonlinear systems by backstepping approach. Inf Sci 369:748–764
Su Y, Wang Q, Sun C (2019) Self-triggered consensus control for linear multi-agent systems with input saturation. IEEE/CAA J Automatica Sinica 7(1):150–157
Sumpter D, Pratt S (2003) A modelling framework for understanding social insect foraging. Behav Ecol Sociobiol 53(3):131–144

Tan X, Cao J, Li X (2018) Consensus of leader-following multiagent systems: A distributed event-triggered impulsive control strategy. IEEE Trans Cybernet 49(3):792–801

Wang X, Li S, Yu X, Yang J (2016) Distributed active anti-disturbance consensus for leader-follower higher-order multi-agent systems with mismatched disturbances. IEEE Trans Autom Control 62(11):5795–5801

Wang A, Liao X, Dong T (2018) Fractional-order follower observer design for tracking consensus in second-order leader multi-agent systems: Periodic sampled-based event-triggered control. J Franklin Inst 355(11):4618–4628

Wang W, Liang H, Pan Y, Li T (2020) Prescribed performance adaptive fuzzy containment control for nonlinear multi-agent systems using disturbance observer. IEEE Trans Cybernet 50(9):3879–3891

Wu H, Su H (2018) Observer-based consensus for positive multiagent systems with directed topology and nonlinear control input. IEEE Trans Syst Man Cybernet Syst 49(7):1459–1469

Wu Y, Wang Z, Ding S, Zhang H (2018) Leader-follower consensus of multi-agent systems in directed networks with actuator faults. Neurocomputing 275:1177–1185

Xie G, Sun L, Wen T, Hei X, Qian F (2021) Adaptive transition probability matrix-based parallel IMM algorithm. IEEE Trans Syst Man Cybernet Syst 51(5):2980–2989

Yao D, Li H, Lu R, Shi Y (2020) Distributed sliding mode tracking control of second-order nonlinear multi-agent systems: an event-triggered approach. IEEE Trans Cybernet 50(9):3892–3902

Ye Y, Su H, Sun Y (2018) Event-triggered consensus tracking for fractional-order multi-agent systems with general linear models. Neurocomputing 315:292–298

Ye F, Sun B, Ou L, Zhang W (2018) Disturbance observer-based control for consensus tracking of multi-agent systems with input delays from a frequency domain perspective. Syst Control Lett 114:66–75

You X, Hua C, Guan X (2018) Event-triggered leader-following consensus for nonlinear multiagent systems subject to actuator saturation using dynamic output feedback method. IEEE Trans Autom Control 63(12):4391–4396

Zhang H, Lewis FL (2012) Adaptive cooperative tracking control of higher-order nonlinear systems with unknown dynamics. Automatica 48(7):1432–1439

Zhang C, Yang G (2018) Event-triggered adaptive output feedback control for a class of uncertain nonlinear systems with actuator failures. IEEE Trans Cybernet 50(1):201–210

Zhang H, Lewis FL, Qu Z (2011) Lyapunov, adaptive, and optimal design techniques for cooperative systems on directed communication graphs. IEEE Trans Indus Electron 59(7):3026–3041

Zhang Y, Liang H, Ma H, Zhou Q, Yu Z (2018) Distributed adaptive consensus tracking control for nonlinear multi-agent systems with state constraints. Appl Math Comput 326:16–32

Zhang XM, Han QL, Ge X, Ding D, Ding L, Yue D, Peng C (2019) Networked control systems: a survey of trends and techniques. IEEE/CAA J Automatica Sinica 7(1):1–17

Zhang L, Liang H, Sun Y, Ahn CK (2021) Adaptive event-triggered fault detection scheme for semi-Markovian jump systems with output quantization. IEEE Trans Syst Man Cybernet Syst 51(4):2370–2381

Zheng S, Shi P, Wang S, Shi Y (2018) Event triggered adaptive fuzzy consensus for interconnected switched multiagent systems. IEEE Trans Fuzzy Syst 27(1):144–158

Zhou Q, Wang W, Liang H, Basin M, Wang B (2021a) Observer-based event-triggered fuzzy adaptive bipartite containment control of multi-agent systems with input quantization. IEEE Trans Fuzzy Syst 29(2):372–384

Zhou Q, Wang W, Ma H, Li H (2021b) Event-triggered fuzzy adaptive containment control for nonlinear multi-agent systems with unknown Bouc-Wen hysteresis input. IEEE Trans Fuzzy Syst 29(4):731–741

Zhou Q, Du P, Li H, Lu R, Yang J (2021c) Adaptive fixed-time control of error-constrained pure-feedback interconnected nonlinear systems. IEEE Trans Syst Man Cybernet Syst 51(10):6369–6380

Zhu S, Liu Y, Lou Y, Cao J (2020) Stabilization of logical control networks: an event-triggered control approach. Sci China Inf Sci 63(1):112203

Zhu Z, Pan Y, Zhou Q, Lu C (2021) Event-triggered adaptive fuzzy control for stochastic nonlinear systems with unmeasured states and unknown backlash-like hysteresis. IEEE Trans Fuzzy Syst 29(5):1273–1283

Chapter 8
Adaptive Fault-Tolerant Containment Control with Input Saturation

Abstract This chapter considers the fault-tolerate containment control problem for stochastic nonlinear multi-agent systems in the presence of input saturation and sensor faults. In order to solve the problem of input saturation, a smooth function is used to approximate the controller saturation function. Due to the excellent approximation characteristic, neural network (NN) is used to deal with unknown nonlinear functions and unknown sensor faults. Meanwhile, by using the auxiliary system and combining adaptive backstepping technique with adaptive NN control, a fault-tolerant containment control approach is proposed. By using graph theory and stochastic Lyapunov stability theory, it can be proved that all signals in the closed-loop system are semi-globally uniformly ultimately bounded in probability. Finally, simulation results are given to show the effectiveness of the proposed control scheme.

8.1 Introduction

During the last decade, the stochastic multi-agent systems (MASs) have obtained great attention due to their applications on unmanned vehicles, unmanned air vehicles, and so on Li and Zhang (2008), Long et al. (2020), Ma et al. (2020), Pan et al. (2022), Liang et al. (2021), Li et al. (2021), Ma et al. (2016), Hua et al. (2017). As one of the most basic control behaviors of MASs, the consensus control has always occupied an extremely important part (Li et al. 2014, 2020; Du et al. 2020; Ren et al. 2020; Pan et al. 2021, 2022; Hu and Feng 2010). The consensus problem can be roughly divided into two categories. One is the leader-following consensus tracking problem (Yao et al. 2020; Tong et al. 2020) and the other is the leaderless consensus problem (Rezaee and Abdollahi 2017; Liang et al. 2021; Zhou et al. 2019). As an extension of the leader-following consensus tracking problem, the containment control is proposed for MASs. In the containment control system, there are multi-leader agents and multi-follower agents (Li et al. 2021). The target of the containment problem is to ensure that all states of followers converge to a convex hull spanned by states of leaders (Wen et al. 2017; Liang et al. 2021). Zhang et al. (2018) applied the containment control to the linear systems with aperiodic sampling. Since then, the containment control has been widely used in various systems, including nonlinear

H. Ren et al., *Intelligent Control of Nonlinear Systems and Nonlinear Multi-Agent Systems with Complex Constraints*, Intelligent Control and Learning Systems 16,
https://doi.org/10.1007/978-981-96-0320-6_8

control systems. Although many researches have been made on containment control, few are oriented towards stochastic nonlinear systems.

As a nonlinear phenomenon, input saturation occurs in many nonlinear MASs, which will influence the stability of systems and limit the performance of systems. Therefore, it is valuable to design controllers for nonlinear MASs with input saturation either practically or theoretically. In view of this, the analysis and synthesis of MASs with input saturation have gotten many attention in recent years (Chen et al. 2015). For example, an observer was used to estimate the external disturbance and the residual term of input saturation in nonlinear nonstrict-feedback MASs (Cao et al. 2021). The observer-based coordinated control protocol was proposed to solve input saturation problem in MASs (Wang et al. 2018). However, the proposed protocol can only be used in undirected topology. By encoding input saturation into optimal control in a reinforcement learning model, the input saturation problem was solved (Qin et al. 2019). The problem of leader-following consensus tracking for linear MASs with input saturation was investigated (Su et al. 2013). Although the input saturation is widely studied in MASs, there are few literature that analyze the MASs with input saturation for containment control problem.

In practical application, sensors of MASs are responsible to transmit data and perceive the outside world (Ren et al. 2020). If the sensor encounters faults, it will influence the security and stability of MASs (Dong et al. 2021; Lin et al. 2022). Therefore, compensating for sensor faults is important. In some literature, sliding mode controllers were employed to detect the sensor faults (Kommuri et al. 2018). Then the finite-time convergence of the error can be ensured by compensating for sensor faults. Meanwhile, to solve the problem of reliability of MASs with sensor faults, an improved adaptive compensation protocol and an improved H_∞ control protocol were proposed, which dealt with sensor faults by using compensators (Chen et al. 2019). An intermediate estimator driven by the relative output measurements was proposed to estimate sensor faults (Zhu et al. 2018). By using a modification technique, the phenomenon of strictly increase in adaptive law caused by sensor faults can be avoided (Deng and Che 2021). A methodology ensured that all agents can reach consensus and the agents can collaborate with other agents to detect sensor faults (Davoodi et al. 2016). Although results related to sensor faults in MASs have been reported, sensor faults for containment control have not been tackled. Therefore, investigating containment control problems with sensor faults in stochastic MASs is still a valuable issue, which is worthy of further research.

Work of Sakthivel et al. (2018) focused on the consensus control of stochastic MASs with input saturation. However, it did not consider the containment control problem of stochastic MASs with the sensor faults. Du et al. (2020) only solved the containment control with stochastic approximation-type methods. It did not take into account sensor faults in stochastic MASs. Motivated by the aforementioned results, this chapter addresses the containment control for nonlinear stochastic MASs with input saturation and sensor faults, which is rarely considered in containment control, by using backstepping technique (Ma et al. 2022).

The main advantages of the proposed control method are listed below:

1. An adaptive neural network (NN) containment control protocol is introduced for stochastic MASs with sensor faults and input saturation. Different from works of Xiong and Hou (2022), Wang et al. (2017), the input saturation and sensor faults are considered in stochastic MASs at the same time.
2. This chapter extends the containment control of stochastic MASs to containment control with input saturation and sensor faults (Su et al. 2020), which makes the proposed control scheme more suitable for engineering applications.
3. For stochastic MASs with input saturation and sensor faults, the stability is analyzed which can ensure all containment control errors in the system are semi-globally uniformly ultimately bounded (SGUUB) in probability.

8.2 Problem Formulation

8.2.1 Systems Description

The ith $(i = 1, \ldots, N)$ follower system is

$$\begin{cases} dx_{im} = (x_{im+1} + f_{im}(\bar{x}_{im}))dt + \psi_{im}^T(\bar{x}_{im})d\omega \\ dx_{in} = (u_i(v_i) + f_{in}(\bar{x}_{in}))dt + \psi_{in}^T(\bar{x}_{in})d\omega \\ \quad y_i = t_i(x_{i1}) \end{cases} \tag{8.1}$$

where $\bar{x}_{im} = [x_{i1}, \ldots, x_{im}]^T \in R^m$ $(m = 1, \ldots, n-1)$ denotes the state vector, and R^m denotes the set of real $m \times 1$ column vectors. $u_i \in R$ is the control input, and $y_i \in R$ is the system output. $f_{ik}(\cdot) \in R^k$ and $\psi_{ik}(\cdot) \in R^k (k = 1, \ldots, n)$ represent the unknown smooth nonlinear functions. $\omega \in R^Q$ means the Q-dimension Brownian process. v_i is the controller output. $u_i(v_i)$ stands for the output of saturation nonlinearity and the system input, which can be expressed by

$$u_i\left(v_i\right) = \begin{cases} \operatorname{sign}\left(v_i\right) u_{B_i}, & |v_i| \geq u_{B_i} \\ v_i, & |v_i| < u_{B_i} \end{cases}$$

where u_{B_i} means the bound of $u_i\left(v_i\right)$. A smooth nonlinear function is used to approximate the saturation function

$$p_i\left(v_i\right) = u_{B_i} \times \tanh\left(\frac{v_i}{u_{B_i}}\right) = u_{B_i} \frac{e^{\left(\frac{v_i}{u_{B_i}}\right)} - e^{\left(-\frac{v_i}{u_{B_i}}\right)}}{e^{\left(\frac{v_i}{u_{B_i}}\right)} + e^{\left(-\frac{v_i}{u_{B_i}}\right)}}$$

Then, $u_i\left(v_i\right)$ can be described as

$$u_i(v_i) = q_i(v_i) + p_i(v_i) = u_{B_i} \times \tanh\left(\frac{v_i}{u_{B_i}}\right) + p_i(v_i)$$

where $p_i(v_i) = u_i(v_i) - q_i(v_i)$ is the bounded function. Its bounds can be represented as

$$|p_i(v_i)| = |u_i(v_i) - q_i(v_i)| \leq u_{B_i}(1 - \tanh(1)) = \rho_i.$$

Thus, system (8.1) can be rewritten as

$$dx_{im} = (x_{im+1} + f_{im}(\bar{x}_{im}))dt + \psi_{im}^T(\bar{x}_{im})d\omega \tag{8.2}$$

$$dx_{in} = (q_i(v_i) + p_i(v_i) + f_{in}(\bar{x}_{in}))dt + \psi_{in}^T(\bar{x}_{in})d\omega \tag{8.3}$$

$$y_i = t_i(x_{i1}). \tag{8.4}$$

According to the work of Cao et al. (2021), the sensor fault model can be expressed as

$$t_i(x_{i1}) = K_i x_{i1} + o_i(t)$$

where $0 < \hbar \leqslant K_i \leqslant 1$, $-\bar{o}_i \leqslant o_i(t) \leqslant \bar{o}_i$, $\hbar$ and $\bar{o}_i$ are constants.

1. When $K_i = 1$, $o_i(\cdot)$ is a positive constant, the bias fault of sensor happens.
2. When $K_i = 1$, $|o_i(\cdot)| = \iota_0 t$ $(0 < \iota_0 \ll 1)$, the drift fault of sensor happens.
3. When $K_i = 1$, $|o_i(\cdot)| < \bar{o}_i$, $o_i(\cdot) \to 0$, it means the sensor is loss of accuracy.
4. When $0 < \hbar \leq K_i < 1$, $o_i(\cdot) = 0$, it indicates that loss of effectiveness of sensor occurs.

Let $f_{swi} = (K_i - 1)x_{i1} + o_i(\cdot)$. y_i can be represented as

$$y_i = x_{i1} + f_{swi}. \tag{8.5}$$

According to (8.5) and (8.1), one has

$$\dot{y}_i = (x_{i2} + f_{i1}(x_{i1}) + f_{isp})dt + K_i\psi_{i1}^T(x_{i1})d\omega \tag{8.6}$$

where $f_{isp} = (K_i - 1)(x_{i2} + f_{i1}(x_{i1})) + \dot{o}_i(\cdot)$.

8.2.2 Graph Theory

For a group of M leaders and a group of N followers, a weighted directed graph $\Gamma = (G, F, H)$ can be used to illustrate the communication relationships, Ren et al. (2020) where $G = \{g_1, g_2 \ldots, g_{N+M}\}$ is the set of nodes, $F = G \times G$ is the set of edges, and $H = [a_{ij}] \in R^{(N+M)\times(N+M)}$ $(i, j = 1, 2, \ldots, N + M)$ is a weighted adjacency matrix. $R^{(N+M)\times(N+M)}$ represents the matrix of $(N+M)\times(N+M)$. If $a_{ij}=1$, the node i can receive the information from node j and otherwise $a_{ij}=0$, $a_{ii}=0$. The Laplacian matrix L is defined as $L=D-H$, where $D=\mathrm{diag}\{d_1,d_2\ldots,d_{N+M}\}\in R^{(N+M)\times(N+M)}$ $(d_i=\Sigma_{j=1}^{N+M}a_{ij})$.

In this chapter, M leader agents and N follower agents will be under the condition of directed topology. Supposed that the followers connect one follower agent at least and that there has no communication between leaders, L can be expressed as

$$L=\begin{bmatrix} L_1 & L_2 \\ 0_{M\times N} & 0_{M\times M} \end{bmatrix}$$

where $L_1\in R^{N\times N}$, $L_2\in R^{N\times M}$.

Assumption 8.1 (Zhou et al. 2021) There is at least one directed path from leader to follower for each follower.

Assumption 8.2 (Wu et al. 2021) Outputs of leader agents y_{Sd} and their time derivatives of signals are bounded, and they satisfied $[y_{Sd},\dot{y}_{Sd},\ldots,y_{Sd}^{(n)}]\in\mathbb{U}_y$, $(||[y_{Sd}||^2+\|\dot{y}_{Sd}\|^2+\cdots+\|y_{Sd}^{(n)}\|^2)\leq\beth_0$, $S\in(N+1,\ldots,N+M)$, where $\beth_0$ is a positive constant.

8.2.3 Radial Basis Function NN

In this chapter, radial basis function NNs are used to approximate unknown nonlinear functions $\psi_{im}^T(\cdot)$ and $f_{im}(\cdot)$. With large enough number of nodes k, over a compact set Ω, the radial basis function NN can approximate any continuous function to arbitrary accuracy, which is described as:

$$f(\chi)=\varsigma^{*T}\Phi(\chi)+\epsilon(\chi)$$

where $f(\chi)$ is a smooth continuous function, $\chi\in\Omega$ is the input vector, and $\epsilon(\chi)$ satisfies $|\epsilon(\chi)|\leq\varepsilon$ $(0<\varepsilon)$. $\Phi(\chi)=[\Phi_1(\chi),\Phi_2(\chi),\ldots,\Phi_k(\chi)]^T$ is the basis function vector, which can be described as

$$\Phi(\chi)=\exp\left[-\frac{(\chi-\eta_i)^T(\chi-\eta_i)}{\mathfrak{W}_i^2}\right], i=1,2,\ldots,k$$

where $\eta_i\in R^n$ denotes the center of the receptive field and $\mathfrak{W}_i\in R^n$ is the width of the Gaussian function. $\varsigma^*=[\varsigma_1^*,\varsigma_2^*,\ldots,\varsigma_\gamma^*]^T\in R^\gamma$ means the ideal weight vector. It can be shown as:

$$\varsigma^*:=\arg\min_{\varsigma\in R^k}\left\{\sup_{\chi\in\Omega}\left|f(\chi)-\varsigma^T\Phi(\chi)\right|\right\}.$$

Lemma 8.1 (Zhou et al. 2021) *Define a continuous function $h(w)$ on compact set Γ. Then one can obtain*

$$\sup_{w\in\Gamma} \left|h(w) - \varsigma^T \varphi(w)\right| \leq \varepsilon, \quad \forall \varepsilon > 0.$$

θ_{ij} is used to express weight vector ς.

$$\theta_{ij} = \|\varsigma_{ij}\|^{\frac{4}{3}}, \quad i = 1, \ldots N, \ j = 1, \ldots, N. \tag{8.7}$$

$\|\cdot\|$ stands for Euclidean norm. Obviously, $\theta_{ij} \geq 0$. We use $\hat{\theta}_{ij}$ to describe the estimation of θ_{ij}. $\tilde{\theta}_{ij} = \theta_{ij} - \hat{\theta}_{ij}$ denotes the estimated error.

Consider the following stochastic system:

$$dz = f(z, t)dt + b(z, t)dy \tag{8.8}$$

where y is an independent r-dimensional Brownian motion and $z \in R^n$ is the state of the system. $b(z, t)$ and $f(z, t)$ are locally Lipschitz functions and satisfy $f(0, t) = b(0, t) = 0$ for $\forall t \geq 0$.

Definition 8.1 (Wu et al. 2021) For a twice continuously differentiable, positive definite and radially unbounded function $R(z, t)$, if it is associated with (8.8), the differential operator ℓ can be calculated as:

$$\ell R = \frac{\partial R}{\partial t} + \frac{\partial R}{\partial z} f + \frac{1}{2} \operatorname{Tr}\left\{b^T \frac{\partial^2 R}{\partial z^2} b\right\} \tag{8.9}$$

where $\operatorname{Tr}(\cdot)$ is the trace of the matrix$(\cdot)$.

The input saturation and senor faults are rarely taken into account together in stochastic MASs. Therefore, in this chapter, the objective is to design a fault-tolerant containment control scheme for the stochastic system (8.1) with input saturation, such that all signals in the closed-loop system are SGUUB in probability.

8.3 Adaptive Containment Controllers Design

In this section, the backstepping technique will be used to design the adaptive controller. First, containment control errors are described as:

$$z_{i1} = \sum_{j=1}^{N} a_{ij}(y_i - y_j) + \sum_{S=N+1}^{N+M} a_{iS}(y_i - y_{Sd}) \tag{8.10}$$

$$z_{im} = x_{i,m} - \alpha_{i,m-1}, m = (2, \ldots, n-1) \tag{8.11}$$

$$z_{in} = x_{i,n} - \alpha_{i,n-1} - \tilde{L}_i \tag{8.12}$$

where $\alpha_{i,m-1}$ is the virtual controller at Step $m-1$ and $\tilde{L}_i$ is an auxiliary control signal.

Step 1: ($m = 1$) Acoording to (8.10), one has

$$\begin{aligned} dz_{i1} &= \sum_{j=1}^{N} a_{ij}\left(\dot{y}_i - \dot{y}_j\right) + \sum_{S=N+1}^{N+M} a_{iS}\left(\dot{y}_i - \dot{y}_{Sd}\right) \\ &= d_i\left(\dot{y}_i - \frac{1}{d_i}\sum_{j=1}^{N} a_{ij}\dot{y}_j - \frac{1}{d_i}\sum_{S=N+1}^{N+M} a_{iS}\dot{y}_{Sd}\right). \end{aligned} \tag{8.13}$$

Substituting (8.1) and (8.6) into (8.13), we have

$$\begin{aligned} dz_{i1} = {} & d_i\left(\dot{y}_i - \frac{1}{d_i}\sum_{j=1}^{N} a_{ij}\dot{y}_j - \frac{1}{d_i}\sum_{S=N+1}^{N+M} a_{i1}\dot{y}_{Sd}\right) \\ = {} & d_i(x_{i2} + f_{i1}(x_{i1}) + f_{isp} - \frac{1}{d_i}\sum_{j=1}^{N} a_{ij}x_{j2} - \frac{1}{d_i}\sum_{j=1}^{N} a_{ij}f_{j1}(x_{j1}) \\ & - \frac{1}{d_i}\sum_{j=1}^{N} a_{ij}f_{jsp} - \frac{1}{d_i}\sum_{S=N+1}^{N+M} a_{iS}\dot{y}_{Sd})dt \\ & + (K_i d_i \psi_{i1}(x_{i1}) - \sum_{j=1}^{N} a_{ij}K_j\psi_{j1}(x_{j1}))d\omega. \end{aligned} \tag{8.14}$$

The Lyapunov candidate is considered as follows

$$V_{i1} = \frac{1}{4}z_{i1}^4 + \frac{1}{2r_{i1}}\tilde{\theta}_{i1}^2 + \frac{1}{2\beta_i}\tilde{\Theta}_i^2 \tag{8.15}$$

where $\beta_i, r_{i1} > 0$ are constants. $\Theta_i = \|\Upsilon_i\|^{\frac{4}{3}}$ and $\theta_{i1} = \|\varsigma_{i1}\|^{\frac{4}{3}}$. According to Lemma 8.1, $\tilde{\theta}_{i1} = \theta_{i1} - \hat{\theta}_{i1}$, $\tilde{\Theta}_i = \Theta_i - \hat{\Theta}_i$. ς_{i1} and Υ_i will be defined later.

By considering Definition 8.1, (8.14), and (8.15), we have

$$\begin{aligned} \ell V_{i1} = {} & z_{i1}^3[d_i(z_{i2} + \alpha_{i,1} + f_{isp} + f_{i1}(x_{i1}) - \frac{1}{d_i}\sum_{j=1}^{N} a_{ij}x_{j2} \\ & - \frac{1}{d_i}\sum_{j=1}^{N} a_{ij}(f_{j1}(x_{j1}) + f_{jsp}) - \frac{1}{d_i}\sum_{S=N+1}^{N+M} a_{iS}\dot{y}_{Sd})] \\ & + \frac{3}{2}z_{i1}^2\Psi_{i1}^T\Psi_{i1} - \frac{1}{r_{i1}}\tilde{\theta}_{i1}\dot{\hat{\theta}}_{i1} - \frac{1}{\beta_i}\tilde{\Theta}_i\dot{\hat{\Theta}}_i \end{aligned} \tag{8.16}$$

where $\Psi_{i1} = d_i K_i \psi_{i1}(x_{i1}) - \sum_{j=1}^{N} a_{ij} K_j \psi_{j1}(x_{j1})$. Based on Young's inequality in work of Cao et al. (2021), we have

$$\frac{3}{2} z_{i1}^2 \Psi_{i1}^T \Psi_{i1} \leq \frac{3}{4 l_{i1}^2} z_{i1}^4 \|\Psi_{i1}\|^4 + \frac{3}{4} l_{i1}^2 \tag{8.17}$$

where l_{i1} is a constant. Substituting (8.17) into (8.16), we have

$$\begin{aligned} \ell V_{i1} \leq\ & z_{i1}^3 \Bigg[d_i (z_{i2} + \alpha_{i,1} + f_{isp} + f_{i1}(x_{i1}) - \frac{1}{d_i} \sum_{j=1}^{N} a_{ij} x_{j2} \\ & - \frac{1}{d_i} \sum_{j=1}^{N} a_{ij} (f_{j1}(x_{j1}) + f_{jsp}) - \frac{1}{d_i} \sum_{S=N+1}^{N+M} a_{iS} \dot{y}_{Sd}) \Bigg] \\ & + \frac{3}{4 l_{i1}^2} z_{i1}^4 \|\Psi_{i1}\|^4 + \frac{3}{4} l_{i1}^2 - \frac{1}{r_{i1}} \tilde{\theta}_{i1} \dot{\hat{\theta}}_{i1} - \frac{1}{\beta_i} \tilde{\Theta}_i \dot{\hat{\Theta}}_i . \end{aligned}$$

Let

$$\begin{aligned} \bar{f}_{i1} &= f_{i1}(x_{i1}) - \frac{1}{d_i} \sum_{j=1}^{N} a_{ij} f_{j1}(x_{j1}) + \frac{3}{4 l_{i1}^2 d_i} z_{i1} \|\Psi_{i1}\|^4 \\ \bar{f}_{isp} &= f_{isp} - \frac{1}{d_i} \sum_{j=1}^{N} a_{ij} f_{jsp}. \end{aligned}$$

We can obtain

$$\begin{aligned} \ell V_{i1} \leq\ & z_{i1}^3 \Bigg[d_i (z_{i2} + \alpha_{i,1} + \bar{f}_{i1} + \bar{f}_{isp} - \frac{1}{d_i} \sum_{j=1}^{N} a_{ij} x_{j2} - \frac{1}{d_i} \sum_{S=N+1}^{N+M} a_{iS} \dot{y}_{Sd}) \Bigg] \\ & + \frac{3}{4} l_{i1}^2 - \frac{1}{r_{i1}} \tilde{\theta}_{i1} \dot{\hat{\theta}}_{i1} - \frac{1}{\beta_i} \tilde{\Theta}_i \dot{\hat{\Theta}}_i . \end{aligned} \tag{8.18}$$

Using Lemma 8.1, we can obtain

$$\bar{f}_{i1} = \varsigma_{i1}^T \Phi_{i1}(Y_{i,1}) + \epsilon(Y_{i,1}) \tag{8.19}$$

where $Y_{i,1} = \{x_{i1}, z_{i1}, a_{ij} x_{j1} (j = 1, \ldots, N)\} \in \Omega_1$ and $|\epsilon(Y_{i,1})| \leq \varepsilon_{i1}$. Ω_1 is the compact set in which the NN can approximate the target function. ς_{i1} is the ideal weight vector of (8.19). Using Young's inequality and considering (8.7), we have

$$\begin{aligned} z_{i1}^3 \bar{f}_{i1} =\ & z_{i1}^3 (\varsigma_{i1}^T \Phi_{i1}(Y_{i,1}) + \epsilon(Y_{i,1})) \\ \leq\ & \frac{3}{4} z_{i1}^4 A_{i1}^{-\frac{4}{3}} \theta_{i1} (\Phi_{i1}^T (Y_{i,1}) \Phi_{i1}(Y_{i,1}))^{\frac{2}{3}} + \frac{3}{4} z_{i1}^4 + \frac{1}{4} A_{i1}^4 + \frac{1}{4} \varepsilon_{i1}^4 . \end{aligned} \tag{8.20}$$

Similarly,

$$\bar{f}_{isp} = \Upsilon_i^T \Phi_{isp}(G_{i,1}) + \epsilon(G_{i,1}) \tag{8.21}$$

where $G_{i,1} = \{x_{i1}, y_i, a_{ij}x_{j1}(j = 1, \ldots, N)\} \in \Omega_{sp}$, and $|\epsilon(G_{i,1})| \leq \xi_i \in R$. Υ_i is the ideal weight vector of (8.21). Then we have

$$z_{i1}^3 \bar{f}_{isp} \leq \frac{3}{4} z_{i1}^4 + \frac{1}{4} b_i^4 + \frac{1}{4} \xi_i^4 + \frac{3}{4} z_{i1}^4 b_i^{-\frac{4}{3}} \Theta_i (\Phi_{isp}^T(G_{i,1}) \Phi_{isp}(G_{i,1}))^{\frac{2}{3}} \tag{8.22}$$

where b_i and A_{i1} are positive constants. Substituting (8.20) and (8.22) into (8.18), we can obtain

$$\begin{aligned}
\ell V_{i1} \leq\ & z_{i1}^3[d_i(z_{i2} + \alpha_{i,1} + \frac{3}{4} z_{i1} b_i^{-\frac{4}{3}} \hat{\Theta}_i (\Phi_{isp}^T(G_{i,1}) \Phi_{isp}(G_{i,1}))^{\frac{2}{3}} \\
& + \frac{3}{4} z_{i1} A_{i1}^{-\frac{4}{3}} \hat{\theta}_{i1} (\Phi_{i1}^T(Y_{i,1}) \Phi_{i1}(Y_{i,1}))^{\frac{2}{3}} - \frac{1}{d_i} \sum_{S=N+1}^{N+M} a_{iS} \dot{y}_{Sd} \\
& - \frac{1}{d_i} \sum_{j=1}^{N} a_{ij} x_{j2} + \frac{3}{2} z_{i1})] + \frac{3}{4} l_{i1}^2 + \frac{d_i}{4} (\xi_i^4 + b_i^4 + A_{i1}^4 + \varepsilon_{i1}^4) \\
& - \frac{1}{r_{i1}} \tilde{\theta}_{i1} (\dot{\hat{\theta}}_{i1} - \frac{3 d_i r_{i1}}{4} A_{i1}^{-\frac{4}{3}} z_{i1}^4 (\Phi_{i1}^T(Y_{i,1}) \Phi_{i1}(Y_{i,1}))^{\frac{2}{3}}) \\
& - \frac{1}{\beta_i} \tilde{\Theta}_i (\dot{\hat{\Theta}}_i - \frac{3 d_i \beta_i}{4} b_i^{-\frac{4}{3}} z_{i1}^4 (\Phi_{isp}^T(G_{i,1}) \Phi_{isp}(G_{i,1}))^{\frac{2}{3}}).
\end{aligned} \tag{8.23}$$

Choose the following virtual controller $\alpha_{i,1}$ as

$$\begin{aligned}
\alpha_{i,1} =& -(k_{i1} + \frac{3}{2}) z_{i1} - \frac{3}{4} z_{i1} A_{i1}^{-\frac{4}{3}} \hat{\theta}_{i1} (\Phi_{i1}^T(Y_{i,1}) \Phi_{i1}(Y_{i,1}))^{\frac{2}{3}} \\
& - \frac{3}{4} z_{i1} b_i^{-\frac{4}{3}} \hat{\Theta}_i (\Phi_{isp}^T(G_{i,1}) \Phi_{isp}(G_{i,1}))^{\frac{2}{3}} \\
& + \frac{1}{d_i} \sum_{S=N+1}^{N+M} a_{iS} \dot{y}_{Sd} + \frac{1}{d_i} \sum_{j=1}^{N} a_{ij} x_{j2}
\end{aligned} \tag{8.24}$$

where k_{i1} is a designed positive parameter. The adaptive law $\dot{\hat{\theta}}_{i1}$ and error adaptive law $\dot{\hat{\Theta}}_i$ are chosen as

$$\dot{\hat{\theta}}_{i1} = -\varrho_{i1} \hat{\theta}_{i1} + \frac{3}{4} d_i r_{i1} A_{i1}^{-\frac{4}{3}} z_{i1}^4 (\Phi_{i1}^T(Y_{i,1}) \Phi_{i1}(Y_{i,1}))^{\frac{2}{3}} \tag{8.25}$$

$$\dot{\hat{\Theta}}_i = -c_i \hat{\Theta}_i + \frac{3}{4} d_i \beta_i b_i^{-\frac{4}{3}} z_{i1}^4 (\Phi_{isp}^T(G_{i,1}) \Phi_{isp}(G_{i,1}))^{\frac{2}{3}} \tag{8.26}$$

where $c_i > 0$ and $\varrho_{i1} > 0$ are designed parameters. Substituting (8.24), (8.25), and (8.26) into (8.23) gets

$$\begin{aligned}\ell V_{i1} \leq\ & z_{i1}^3 d_i (z_{i2} - k_{i1} z_{i1}) + \frac{3}{4} l_{i1}^2 + \frac{d_i}{4} (\xi_i^4 + b_i^4 + A_{i1}^4 + \varepsilon_{i1}^4) \\ & + \frac{\varrho_{i1}}{r_{i1}} \tilde{\theta}_{i1} \hat{\theta}_{i1} + \frac{c_i}{\beta_i} \tilde{\Theta}_i \hat{\Theta}_i . \end{aligned} \tag{8.27}$$

Step 2: ($m = 2$) Differentiating (8.11) with respect to time and considering (8.1), we have

$$dz_{i2} = (x_{i3} + f_{i2}(\bar{x}_{i2}) - \ell \alpha_{i,1}) dt + (\psi_{i2} - \frac{\partial \alpha_{i,1}}{\partial x_{i1}} K_i \psi_{i1})^T d\omega \tag{8.28}$$

with

$$\begin{aligned}\ell \alpha_{i,1} =\ & \frac{\partial \alpha_{i,1}}{\partial x_{i1}} \left[x_{i2} + f_{i1} \left(\bar{x}_{i1} \right) \right] + \frac{\partial \alpha_{i,1}}{\partial \hat{\theta}_{i1}} \dot{\hat{\theta}}_{i1} + \sum_{r=0}^{1} \sum_{l=N+1}^{N+M} \frac{\partial \alpha_{i,1}}{\partial y_{ld}^r} y_{ld}^{(r)} \\ & + \frac{1}{2} \frac{\partial^2 \alpha_{i,1}}{\partial x_{i1}^2} (K_i \psi_{i1})^T (K_i \psi_{i1}). \end{aligned}$$

Similarly to (8.15), $\tilde{\theta}_{i2} = \theta_{i2} - \hat{\theta}_{i2}$. Choose the following Lyapunon function

$$V_{i2} = V_{i1} + \frac{1}{4} z_{i2}^4 + \frac{1}{2 r_{i2}} \tilde{\theta}_{i2}^2 \tag{8.29}$$

where $r_{i2} > 0$ is a designed parameter. Similarly to (8.16), it follows

$$\ell V_{i2} \leq \ell V_{i1} + z_{i2}(z_{i3} + \alpha_{i,2} + f_{i2}(\bar{x}_{i2}) - \ell \alpha_{i,1}) + \frac{3}{2} z_{i2}^2 \Psi_{i2}^T \Psi_{i2} - \frac{1}{r_{i2}} \tilde{\theta}_{i2} \dot{\hat{\theta}}_{i2} \tag{8.30}$$

where $\Psi_{i2} = \psi_{i2} - \frac{\partial \alpha_{i,1}}{\partial x_{i1}} K_i \psi_{i1}$. Similarly to (8.17), we have

$$\frac{3}{2} z_{i2}^2 \Psi_{i2}^T \Psi_{i2} \leq \frac{3}{4 l_{i2}^2} z_{i2}^4 \| \Psi_{i2} \|^4 + \frac{3}{4} l_{i2}^2 \tag{8.31}$$

where l_{i2} is a constant. Substituting (8.31) into (8.30), we have

$$\ell V_{i2} = \ell V_{i1} + z_{i2}^3 \left(z_{i3} + \alpha_{i,2} + \bar{f}_{i2} \right) + \frac{3}{4} l_{i2}^2 - \frac{1}{\gamma_{i2}} \tilde{\theta}_{i2} \dot{\hat{\theta}}_{i2}$$

where

$$\bar{f}_{i2} = f_{i2} \left(\bar{x}_{i2} \right) + \frac{3}{4 l_{i2}^2} z_{i2} \left\| \Psi_{i2} \right\|^4 - \ell \alpha_{i,1} .$$

Similarly to (8.20), we have

$$z_{i2}^{3}\bar{f}_{i2} \leq \frac{3}{4}A_{i2}^{-\frac{4}{3}}z_{i2}^{4}\theta_{i2}(\Phi_{i2}^{T}(Y_{i,2})\Phi_{i2}(Y_{i,2}))^{\frac{2}{3}} + \frac{1}{4}A_{i2}^{4} + \frac{3}{4}z_{i2}^{4} + \frac{1}{4}\varepsilon_{i2}^{4}$$

where A_{i2} is a positive constant. $Y_{i,2} = \{\bar{x}_{i2}, z_{i2}, y_{Sd}^{r}, \dot{\hat{\theta}}_{i1}, \dot{\hat{\Theta}}_{i}, \hat{\theta}_{i1}, \hat{\Theta}_{i}\} \in \Omega_2$. Similarly to (8.23), we have

$$\begin{aligned}\ell V_{i2} \leq\ & \ell V_{i1} + z_{i2}^{3}\Big(z_{i3} + \frac{3}{4}A_{i2}^{-\frac{4}{3}}z_{i2}\hat{\theta}_{i2}(\Phi_{i2}^{T}(Y_{i,2})\Phi_{i2}(Y_{i,2}))^{\frac{2}{3}} + \alpha_{i,2} + \frac{3}{4}z_{i2}\Big)\\ &+ \frac{3}{4}l_{i2}^{2} + \frac{1}{4}(A_{i2}^{4} + \varepsilon_{i2}^{4})\\ &- \frac{\tilde{\theta}_{i2}}{r_{i2}}\Big(\dot{\hat{\theta}}_{i2} - \frac{3r_{i2}}{4}A_{i2}^{-\frac{4}{3}}z_{i2}^{4}(\Phi_{i2}^{T}(Y_{i,2})\Phi_{i2}(Y_{i,2}))^{\frac{2}{3}}\Big).\end{aligned} \tag{8.32}$$

The virtual controller $\alpha_{i,2}$ is designed as

$$\alpha_{i,2} = -(k_{i2} + \frac{3}{4})z_{i2} - \frac{3}{4}A_{i2}^{-\frac{4}{3}}\hat{\theta}_{i2}z_{i2}(\Phi_{i2}^{T}(Y_{i,2})\Phi_{i2}(Y_{i,2}))^{\frac{2}{3}} \tag{8.33}$$

where k_{i2} is a designed parameter. The adaptive law $\dot{\hat{\theta}}_{i2}$ is given as

$$\dot{\hat{\theta}}_{i2} = -\varrho_{i2}\hat{\theta}_{i2} + \frac{3}{4}r_{i2}A_{i2}^{-\frac{4}{3}}z_{i2}^{4}(\Phi_{i2}^{T}(Y_{i,2})\Phi_{i2}(Y_{i,2}))^{\frac{2}{3}} \tag{8.34}$$

where $\varrho_{i2} > 0$ is a designed parameter. Substituting (8.33) and (8.34) into (8.32), we have

$$\ell V_{i2} \leq \ell V_{i1} + z_{i2}^{3}(z_{i3} - k_{i2}z_{i2}) + \frac{3}{4}l_{i2}^{2} + \frac{1}{4}(A_{i2}^{4} + \varepsilon_{i2}^{4}) + \frac{\varrho_{i2}}{r_{i2}}\tilde{\theta}_{i2}\hat{\theta}_{i2}.$$

Step m**:** $(3 \leq m \leq n-1)$ Considering (8.1) and differentiating (8.11) with respect to time, we have

$$dz_i = (x_{im+1} + f_{im}(\bar{x}_{im}) - \ell\alpha_{i,m-1})dt + (\psi_{im} - \sum_{w=1}^{m-1}\frac{\partial\alpha_{i,m-1}}{\partial x_{iw}}\psi_{iw})^{T}d\omega$$

with

$$\begin{aligned}\ell\alpha_{i,m-1} =& \sum_{r=1}^{m-1}\frac{\partial\alpha_{i,m-1}}{\partial x_{ir}}\left[x_{ir+1} + f_{ir}(\bar{x}_{ir})\right] + \sum_{r=1}^{m-1}\frac{\partial\alpha_{i,m-1}}{\partial\hat{\theta}_{ir}}\dot{\hat{\theta}}_{ir} +\\ &\sum_{r=0}^{m-1}\sum_{l=N+1}^{N+M}\frac{\partial\alpha_{i,m-1}}{\partial y_{ld}^{r}}y_{ld}^{(r+1)} + \frac{1}{2}\sum_{p,q=1}^{m-1}\frac{\partial^{2}\alpha_{i,m-1}}{\partial x_{ip}\partial x_{iq}}(J_p\psi_{ip})^{T}(J_q\psi_{iq}).\end{aligned} \tag{8.35}$$

In (8.35),

$$J_p = \begin{cases} K_i, & p = 1 \\ 1, & p \neq 1 \end{cases}$$

and

$$J_q = \begin{cases} K_i, & q = 1 \\ 1, & q \neq 1 \end{cases}.$$

Similarly to (8.15), $\tilde{\theta}_{im} = \theta_{im} - \hat{\theta}_{im}$. Choose the following Lyapunon function

$$V_{im} = V_{im-1} + \frac{1}{4}z_{im}^4 + \frac{1}{2r_{im}}\tilde{\theta}_{im}^2$$

where $r_{im} > 0$ is a designed parameter. Similarly to (8.30), it follows

$$\begin{aligned} \ell V_{im} \leq\ & \ell V_{im-1} + z_{im}(z_{im+1} + \alpha_{i,m} + f_{im}(\bar{x}_{im}) - \ell\alpha_{i,m-1}) \\ & + \frac{3}{2}z_{im}^2\Psi_{im}^T\Psi_{im} - \frac{1}{r_{im}}\tilde{\theta}_{im}\dot{\hat{\theta}}_{im} \end{aligned} \tag{8.36}$$

where

$$\Psi_{im} = \psi_{im} - \sum_{w=2}^{m-1}\frac{\partial\alpha_{i,m-1}}{\partial x_{iw}}\psi_{iw} - K_i\frac{\partial\alpha_{i,m-1}}{\partial x_{i1}}\psi_{i1}. \tag{8.37}$$

Similarly to (8.17), we have

$$\frac{3}{2}z_{im}^2\Psi_{im}^T\Psi_{im} \leq \frac{3}{4l_{im}^2}z_{im}^4\|\Psi_{im}\|^4 + \frac{3}{4}l_{im}^2 \tag{8.38}$$

where l_{im} is a constant. Substituting (8.38) into (8.36), we have

$$\ell V_{im} \leq \ell V_{im-1} + z_{im}^3\left(z_{im+1} + \alpha_{i,m} + \bar{f}_{im}\right) + \frac{3}{4}l_{im}^2 - \frac{1}{\gamma_{im}}\tilde{\theta}_{im}\dot{\hat{\theta}}_{im} \tag{8.39}$$

where

$$\bar{f}_{im} = f_{im}(\bar{x}_{im}) + \frac{3}{4l_{im}^2}z_{im}\|\Psi_{im}\|^4 - \ell\alpha_{i,m-1}$$

Similarly to (8.20), we have

$$z_{im}^3\bar{f}_{im} \leq \frac{3}{4}A_{im}^{-\frac{4}{3}}z_{im}^4\theta_{im}(\Phi_{im}^T(Y_{i,m})\Phi_{im}(Y_{i,m}))^{\frac{2}{3}} + \frac{1}{4}A_{im}^4 + \frac{3}{4}z_{im}^4 + \frac{1}{4}\varepsilon_{im}^4 \tag{8.40}$$

where A_{im} is a positive constant. According to (8.35), $Y_{i,m} = \{\bar{x}_{im}, z_{im}, y_{Sd}^r, \dot{\hat{\theta}}_{iw-1}, \hat{\theta}_{iw-1} \ (w, r = 1, \ldots, m)\} \in \Omega_m$. Substituting (8.40) into (8.39), we have

$$\ell V_{im} \leq \ell V_{im-1} + z_{im}^3 \left(z_{im-1} + \alpha_{i,m} + \frac{3}{4} A_{im}^{-\frac{4}{3}} z_{im} \hat{\theta}_{im} (\Phi_{im}^T(Y_{i,m}) \Phi_{im}(Y_{i,m}))^{\frac{2}{3}} + \frac{3}{4} z_{im} \right) - \frac{\tilde{\theta}_{im}}{r_{im}} \left(\dot{\hat{\theta}}_{im} - \frac{3 r_{im}}{4} A_{im}^{-\frac{4}{3}} z_{im}^4 (\Phi_{im}^T(Y_{i,m}) \Phi_{im}(Y_{i,m}))^{\frac{2}{3}} \right) + \frac{3}{4} l_{im}^2 + \frac{1}{4} (A_{im}^4 + \varepsilon_{im}^4). \tag{8.41}$$

The virtual controller $\alpha_{i,m}$ is designed as

$$\alpha_{i,m} = -\left(k_{im} + \frac{3}{4} \right) z_{im} - \frac{3}{4} A_{im}^{-\frac{4}{3}} \hat{\theta}_{im} z_{im} (\Phi_{im}^T(Y_{i,m}) \Phi_{im}(Y_{i,m}))^{\frac{2}{3}} \tag{8.42}$$

where k_{im} is a designed parameter. The adaptive law $\dot{\hat{\theta}}_{im}$ is given as

$$\dot{\hat{\theta}}_{im} = -\varrho_{im} \hat{\theta}_{im} + \frac{3}{4} r_{im} A_{im}^{-\frac{4}{3}} z_{im}^4 (\Phi_{im}^T(Y_{i,m}) \Phi_{im}(Y_{i,m}))^{\frac{2}{3}} \tag{8.43}$$

where $\varrho_{im} > 0$ is a designed parameter. Substituting (8.42) and (8.43) into (8.41), we have

$$\ell V_{im} \leq \ell V_{im-1} + z_{im}^3 (z_{im+1} - k_{im} z_{im}) + \frac{3}{4} l_{im}^2 + \frac{1}{4} (A_{im}^4 + \varepsilon_{im}^4) + \frac{\varrho_{im}}{r_{im}} \tilde{\theta}_{im} \hat{\theta}_{im}.$$

Step n**:** The time derivative of (8.12) is

$$dz_{in} = \left(u(v_i) + f_{in}(\bar{x}_{in}) - \ell \alpha_{i,n-1} + \dot{\hat{L}}_i \right) dt + \left(\psi_{in} - \sum_{w=1}^{n-1} \frac{\partial \alpha_{i,n-1}}{\partial x_{iw}} \psi_{iw} \right)^T d\omega.$$

Define the Lyapunov function as

$$V_{in} = V_{in-1} + \frac{1}{4} z_{in}^4 + \frac{1}{2 r_{in}} \tilde{\theta}_{in}^2 \tag{8.44}$$

where $r_{in} > 0$ is a designed parameter. Taking the time derivative of (8.44), we have

$$\ell V_{in} \leq \ell V_{in-1} + z_{in}^3 (p_i(v_i) + q_i(v_i) + f_{in}(\bar{x}_{in}) - \ell \alpha_{i,n-1} - \dot{\hat{L}}_i) + \frac{3}{2} z_{in}^2 \Psi_{in}^T \Psi_{in} - \frac{1}{r_{in}} \tilde{\theta}_{in} \dot{\hat{\theta}}_{in}. \tag{8.45}$$

According to Young's inequality, one has

$$z_{in}^3 p_i(v_i) \leq \frac{3}{4} z_{in}^4 + \frac{1}{4} \rho_i^4$$

Similarly to (8.17), we have

$$\frac{3}{2} z_{in}^2 \Psi_{in}^T \Psi_{in} \leq \frac{3}{4 l_{in}^2} z_{in}^4 \|\Psi_{in}\|^4 + \frac{3}{4} l_{in}^2$$

where l_{in} is a constant. ℓV_{in} can be rewritten as

$$\ell V_{in} \leq \ell V_{in-1} + z_{in}^3 (q_i(v_i) + \bar{f}_{in} + \frac{3}{4} z_{in} - \dot{\tilde{L}}) + \frac{3}{4} l_{in}^2 + \frac{1}{4} \rho_i^4 - \frac{1}{r_{in}} \tilde{\theta}_{in} \dot{\hat{\theta}}_{in} \quad (8.46)$$

where

$$\bar{f}_{in} = f_{in}(\bar{x}_{in}) + \frac{3}{4 l_{in}^2} z_{in} \|\Psi_{in}\|^4 - \ell \alpha_{i,n-1}.$$

We design the auxiliary controller as

$$\dot{\tilde{L}}_i = -\tilde{L}_i + q_i(v_i) - v_i. \quad (8.47)$$

Remark 8.1 Su et al. (2013) proposed the low gain feedback method to solve the input saturation in MASs. However, he did not consider the stochastic MASs with input saturation and the low gain feedback method only can be used in linear systems. In this chapter, an auxiliary controller is utilized in stochastic MASs to compensate the error caused by input saturation, which is used in most nonlinear MASs rather than only the linear MASs.

Similarly to (8.20), we have

$$z_{in}^3 \bar{f}_{in} \leq \frac{3}{4} A_{in}^{-\frac{4}{3}} z_{in}^4 \theta_{in} (\Phi_{in}^T (Y_{i,n}) \Phi_{in}(Y_{i,n}))^{\frac{2}{3}} + \frac{1}{4} A_{in}^4 + \frac{3}{4} z_{in}^4 + \frac{1}{4} \varepsilon_{in}^4 \quad (8.48)$$

where A_{in} is a positive constant. Substituting (8.47) and (8.48) into (8.46), we have

$$\begin{aligned} \ell V_{in} \leq\ & \ell V_{in-1} + z_{in}^3 \left(\frac{3}{4} A_{in}^{-\frac{4}{3}} z_{in} \theta_{in} (\Phi_{in}^T (Y_{i,n}) \Phi_{in}(Y_{i,n}))^{\frac{2}{3}} + \frac{3}{2} z_{in} + v_i + \tilde{L}_i \right) \\ & + \frac{3}{4} l_{in}^2 - \frac{1}{r_{in}} \tilde{\theta}_{in} \dot{\hat{\theta}}_{in} + \frac{1}{4} (A_{in}^4 + \varepsilon_{in}^4 + \rho_i^4). \end{aligned} \quad (8.49)$$

The adaptive law $\dot{\hat{\theta}}_{in}$ is given as

$$\dot{\hat{\theta}}_{in} = -\varrho_{in} \hat{\theta}_{in} + \frac{3}{4} r_{in} A_{in}^{-\frac{4}{3}} z_{in}^4 (\Phi_{in}^T (Y_{i,n}) \Phi_{in}(Y_{i,n}))^{\frac{2}{3}} \quad (8.50)$$

where $\varrho_{in} > 0$ is a designed parameter. Taking (8.50) into (8.49), we have

$$\begin{aligned}\ell V_{in} \leq\ & \ell V_{in-1} + z_{in}^3\left(\frac{3}{4}A_{in}^{-\frac{4}{3}} z_{in}\theta_{in}(\Phi_{in}^T(Y_{i,n})\Phi_{in}(Y_{i,n}))^{\frac{2}{3}} + \frac{3}{2}z_{in} + v_i + \tilde{L}_i\right)\\ & + \frac{3}{4}l_{in}^2 - \frac{1}{r_{in}}\tilde{\theta}_{in}\dot{\hat{\theta}}_{in} + \frac{1}{4}(A_{in}^4 + \varepsilon_{in}^4 + \rho_i^4). \end{aligned} \tag{8.51}$$

We design the ith input as

$$v_i = -\left(k_{in} + \frac{3}{2}\right)z_{in} - \frac{3}{4}A_{in}^{-\frac{4}{3}} z_{in}\hat{\theta}_{in}(\Phi_{in}^T(Y_{i,n})\Phi_{in}(Y_{i,n}))^{\frac{2}{3}} - \tilde{L}_i. \tag{8.52}$$

Then (8.51) can be rewritten as

$$\ell V_{in} \leq \ell V_{in-1} - k_{in}z_{in}^4 + \frac{1}{4}(A_{in}^4 + \varepsilon_{in}^4 + \rho_i^4) + \frac{\varrho_{in}}{r_{in}}\tilde{\theta}_{in}\hat{\theta}_{in} + \frac{3}{4}l_{in}^2. \tag{8.53}$$

8.4 Stability Analysis

Based on the above discussion, we can obtain the following theorem.

Theorem 8.1 *Consider the system* (8.1) *with the directed graph and Assumption 8.1. Based on* (8.10), (8.11), *and* (8.12), *a fault-tolerate containment controller is proposed to ensure that the containment errors are bounded, which can be described as*

$$E[V(t)] \leq \daleth \tag{8.54}$$

where $\daleth$ *is a positive constant and* $E(\cdot)$ *denotes the mathematical expectation. All signals in the stochastic MASs are SGUUB in probability.*

Proof Define the following Lyapunov function candidate

$$V = \sum_{i=1}^{N} V_{in}. \tag{8.55}$$

According to Young's inequality, ℓV_{in} can be rewritten as

$$\ell V_{in} \leq -z_{i1}^4\left(k_{i1}d_i - \frac{3d_i^4}{4}\right) - \sum_{w=2}^{n-1}(k_{iw}-1)z_{iw}^4 - \left(k_{in}-\frac{1}{4}\right)z_{in}^4 - \sum_{w=1}^{n}\frac{\varrho_{iw}}{2r_{iw}}\tilde{\theta}_{iw}^2 - \frac{c_i}{2\beta_i}\tilde{\Theta}_i^2 + \Delta_i \tag{8.56}$$

where

$$\Delta_i = \frac{d_i}{4}(\xi_i^4 + b_i^4 + A_{i1}^4 + \varepsilon_{i1}^4) + \sum_{w=2}^{n}\left(\frac{3}{4}l_{iw}^2 + \frac{1}{4}(A_{iw}^4 + \varepsilon_{iw}^4)\right) + \frac{1}{4}\rho_i^4 + \frac{c_i}{2\beta_i}\Theta_i^2 + \sum_{w=1}^{n}\frac{\varrho_{iw}}{2r_{iw}}\theta_{iw}^2. \tag{8.57}$$

Inequality (8.56) can be rewritten as

$$\ell V \leq -\sum_{i=1}^{N}\left(k_{i1}d_i - \frac{3d_i^4}{4}\right)z_{i1}^4 - \sum_{i=1}^{N}\sum_{w=2}^{n-1}(k_{iw}-1)z_{iw}^4 - \sum_{i=1}^{N}\left(k_{in}-\frac{1}{4}\right)z_{in}^4 - \sum_{i=1}^{N}\sum_{w=1}^{n}\frac{\varrho_{iw}}{2r_{iw}}\tilde{\theta}_{iw}^2 - \sum_{i=1}^{N}\frac{c_i}{2\beta_i}\tilde{\Theta}_i^2 + \sum_{i=1}^{N}\Delta_i. \tag{8.58}$$

Let

$$B_i = \min_{1\leq i\leq N, 2\leq w\leq n, 1\leq j\leq n}\left\{k_{i1} - \frac{3d_i^3}{4}, k_{iw}-1, k_{in}-\frac{1}{4}, \frac{c_i}{2\beta_i}, \frac{\varrho_{ij}}{2r_{ij}}\right\} \tag{8.59}$$

$$\Delta = \sum_{i=1}^{N}\Delta_i.$$

According to (8.59), $k_{i1} > \frac{3d_i^3}{4}$, $k_{iw} > 1$, $k_{in} - \frac{1}{4}$.

Remark 8.2 According to (8.59), B_i should be as big as possible and Δ is as small as possible. But both them are decided by $\frac{c_i}{2\beta_i}, \frac{\varrho_{iw}}{2r_{iw}}$. Therefore, there are two case, $\max\{k_{i1} - \frac{3d_i^4}{4}, k_{iw}-1, k_{in}-\frac{1}{4}\} < \min\{\frac{c_i}{2\beta_i}, \frac{\varrho_{iw}}{2r_{iw}}\}$ and $\min\{k_{i1} - \frac{3d_i^4}{4}, k_{iw}-1, k_{in}-\frac{1}{4}\} > \max\{\frac{c_i}{2\beta_i}, \frac{\varrho_{iw}}{2r_{iw}}\}$. In case one , we should make $\max\limits_{1\leq i\leq N, 2\leq w\leq n}\{k_{i1} - \frac{3d_i^4}{4}, k_{iw}-1, k_{in}-\frac{1}{4}\} = \min\{\frac{c_i}{2\beta_i}, \frac{\varrho_{iw}}{2r_{iw}}\}$. In case two, we should change $\min\{\frac{c_i}{2\beta_i}, \frac{\varrho_{iw}}{2r_{iw}}\}$ according to the result of simulation.

Therefore, we have

$$\ell V = -B_i V + \Delta, \quad t > 0. \tag{8.60}$$

According to the work of Tonge et al. (2011), for all $x \in R^n$ and $t > t_0$, there is an unique strong solution of system (8.8) for each $x \in R^n$ and it has

$$E[V(t)] \leq V(0)e^{-B_i t} + \frac{\Delta}{B_i}. \tag{8.61}$$

According to (8.61), $E[V(t)] \leq V(0) + \frac{\Delta}{B_i}$. Let $\daleth = V(0) + \frac{\Delta}{B_i}$, we can get

$$E[V(t)] \leq \daleth.$$

Therefore, all signals in the stochastic MASs are SGUUB in probability. The proof is completed. □

Remark 8.3 Different from the work of Li and Mu (2019), this chapter considers sensor faults in containment control of stochastic MASs. Errors of sensor faults are taken account into in the system and Lyapunov function. By using the proposed control scheme, errors caused by sensor faults converge to a fixed constant range and stochastic MASs with sensor faults are stable.

8.5 Numerical Example

In this section, the effectiveness of the containment control scheme is proved by a simulation example. A stochastic nonlinear MAS consists of four followers and two leaders. The system of each follower is modeled by

$$\begin{cases} dx_{i1} = (x_{i2} - 0.2x_{i1} + \sin(x_{i1}))dt + (0.15\cos(x_{i1}))^T d\omega \\ dx_{i2} = (v_i + 0.1\cos(x_{i2}))dt + (0.01x_{i1}^2 \sin(x_{i2}))^T d\omega \\ \quad y_i = K_i x_{i1} + \rho_i(t) \end{cases} \tag{8.62}$$

where y_i is the output of the system. In addition, the outputs of leaders are

$$\begin{aligned} y_{r1} &= 0.5\sin(t) \\ y_{r2} &= 0.5\sin(t) + 1.5. \end{aligned}$$

Some relative parameters are chosen, where $k_{i1} = 12.9$, $k_{i2} = 3.8$, $A_{i1} = 3.8$, $A_{i2} = 68.8$, $\beta_i = 1.05$, $\varrho_{i1} = 0.1$, $\varrho_{i2} = 0.2$, $r_{i1} = 1.05$, $r_{i2} = 2.2$, $b_i = 33.3$, $c_i = 0.1$. The directed graph is shown in Fig. 8.1, where nodes 5 and 6 represent the leader, and outputs of them are y_{Sd1}, y_{Sd2}, respectively. Upper and lower bounds of input of the system are 1.6 and -1.6, respectively. Sensor faults are set to occur in the $20th$ second.

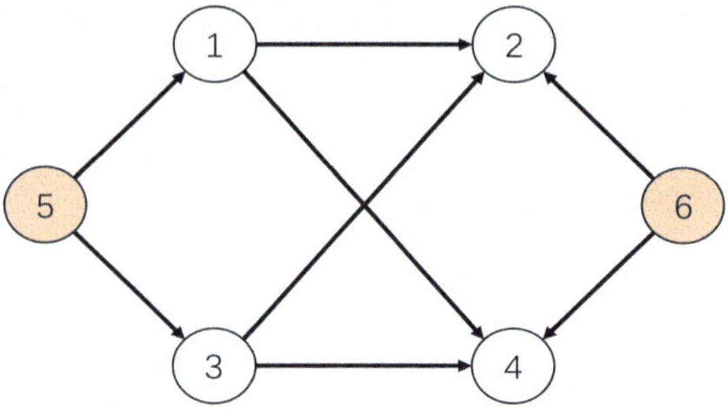

Fig. 8.1 Communication graph

The Laplacian matrix L is given as

$$L = \begin{bmatrix} 3 & -1 & 0 & -1 & -1 & 0 \\ -1 & 3 & -1 & 0 & 0 & -1 \\ -1 & 0 & 2 & -1 & -1 & 0 \\ 0 & 0 & -1 & 2 & 0 & -1 \\ 0 & 0 & 0 & 0 & 0 & 0 \\ 0 & 0 & 0 & 0 & 0 & 0 \end{bmatrix}.$$

For analyzing different types of sensor faults, three groups of parameters are set. The first case only analyzes the loss of effectiveness of sensor. The case 2 analyzes both loss of effectiveness of sensor and bias fault of sensor, which happen at the same time. The case 3 analyzes loss of effectiveness of sensor, bias fault of sensor, and drift fault of sensor, which happen simultaneously.

Case 1: Parameters of sensor faults are $K_i = 0.86$, $\rho_i = 0$. In Fig. 8.2, the signals of followers are contained by the signals of leaders. Obviously, the followers outputs $y_i (i = 1, 2, 3, 4)$ can follow signals of leaders after sensor faults happened, which

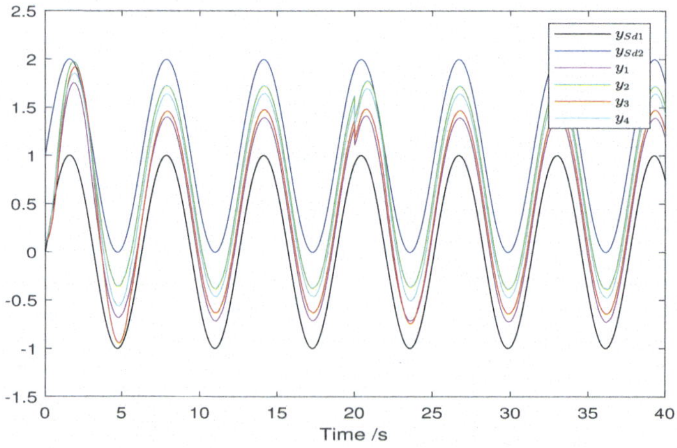

Fig. 8.2 The trajectories of followers y_1, y_2, y_3, y_4, and leaders y_{Sd1}, y_{Sd2} (case 1)

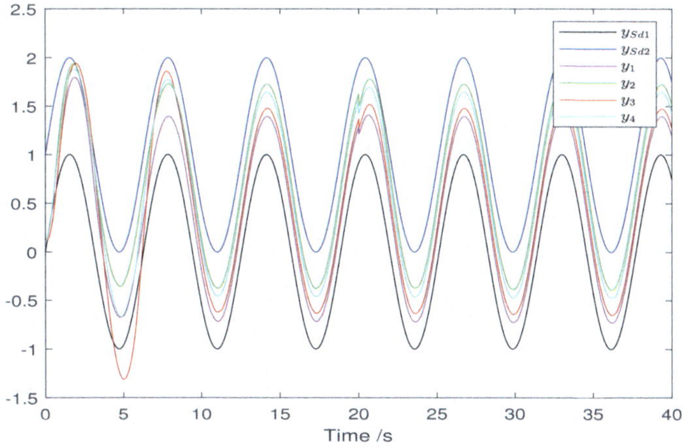

Fig. 8.3 The trajectories of followers y_1, y_2, y_3, y_4, and leaders y_{Sd1}, y_{Sd2} (case 2)

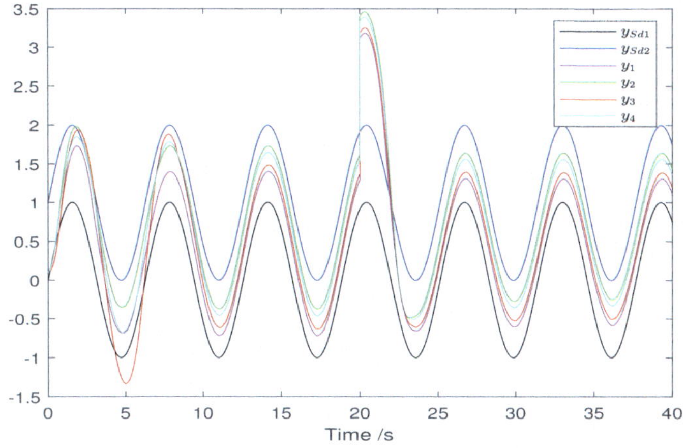

Fig. 8.4 The trajectories of followers y_1, y_2, y_3, y_4, and leaders y_{Sd1}, y_{Sd2} (case 3)

proves that the proposed fault-tolerate containment control scheme is effective for loss of effectiveness of sensor fault.

Case 2: Parameters of sensor faults are $K_i = 0.86$, $\rho_i = 0.1$. Figure 8.3 shows that signs of followers can be contained by signs of leaders, which proves that the proposed fault-tolerate containment control scheme is effective for both loss of effectiveness fault and bias fault of sensor.

Case 3: Parameters of sensor faults are $K_i = 0.86$, $\rho_i = 1 + 0.05t$, which means that loss of effectiveness fault, drift fault and bias fault of sensor occur simultaneously. Figure 8.4 shows that signs of followers suddenly exceed the containment of leaders' signs and the system returns to steady state after few second. It can be concluded

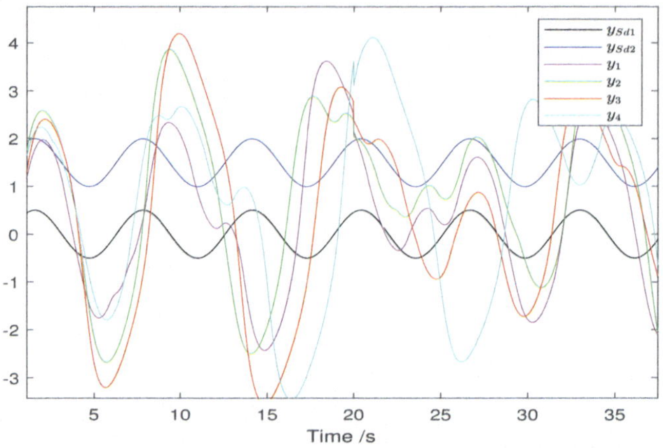

Fig. 8.5 The trajectories of followers without considering the input saturation (followers y_1, y_2, y_3, y_4 and leaders y_{Sd1}, y_{Sd2})

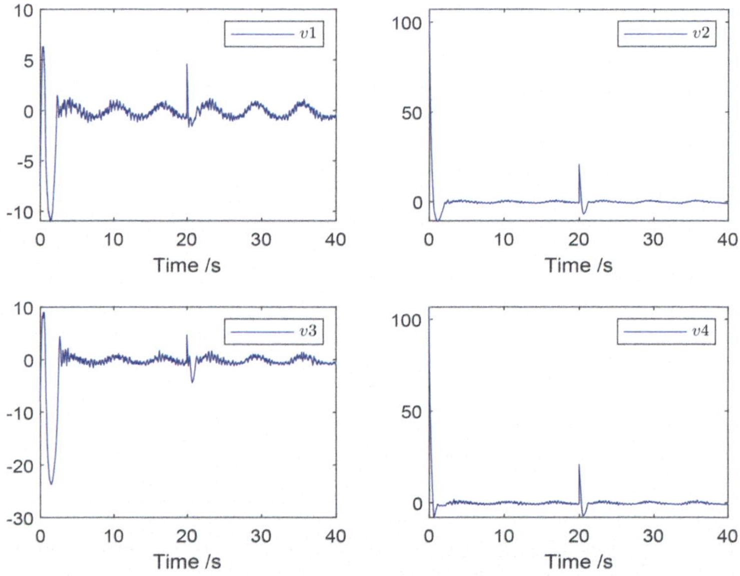

Fig. 8.6 Controller outputs $v_i (i = 1, 2, 3, 4)$ (case 1)

that the control scheme in this chapter can handle the multiple sensor faults at the same time.

Figure 8.5 shows the system output when input saturation is not taken into account. From the figure, trajectories of followers cannot track the leader outputs. When the system has input saturation, the actual input amplitude of the system cannot exceed the maximum allowable input. Therefore, it is impossible to obtain an ideal control

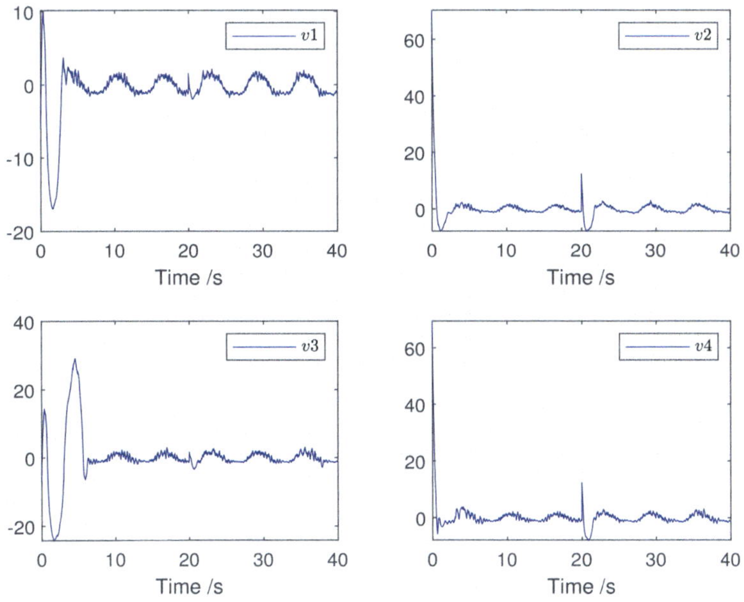

Fig. 8.7 Controller outputs $v_i (i = 1, 2, 3, 4)$ (case 2)

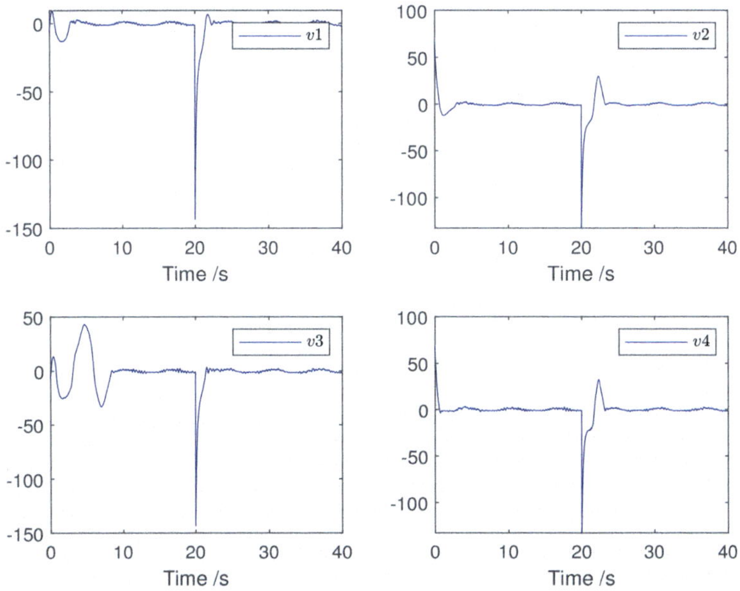

Fig. 8.8 Controller outputs $v_i (i = 1, 2, 3, 4)$ (case 3)

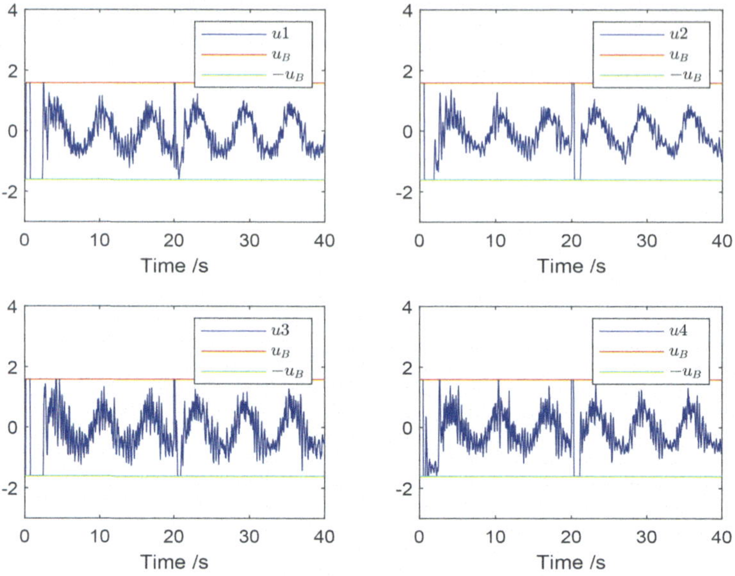

Fig. 8.9 Inputs of followers $u_i (i = 1, 2, 3, 4)$ (case 1)

signal that can achieve system performance. Taking Fig. 8.5 in this chapter as an example, it can be seen that the leader's signal is sinusoidal wave. And waveforms of followers 1 and 3 are roughly sinusoidal waves. But the phases of them are obviously lagging, and their signals exceed the leader's convex hull.

In Figs. 8.9, 8.10, and 8.11, the system inputs u_i $(i = 1, 2, 3, 4)$ of these followers are bounded and u_i are within the upper and lower bounds of the system inputs. The red line and the green line are the upper bound and lower bound of input, respectively. Moreover, Figs. 8.6, 8.7, and 8.8 show that control outputs $v_i (i = 1, 2, 3, 4)$ exceed upper and lower bounds of system input. According to the above pictures, it is proved that the proposed control method can handle the problem of input saturation.

8.6 Conclusions

In this chapter, we have considered containment control problem for stochastic nonlinear MASs in the presence of input saturation and sensor faults. The containment controller can be obtained by using backstepping technique, which can ensure all followers converge to a convex hull spanned by leaders. The auxiliary function and NN have guaranteed that sensor faults and input saturation can be compensated. The effectiveness of the proposed results has been illustrated by the simulation example.

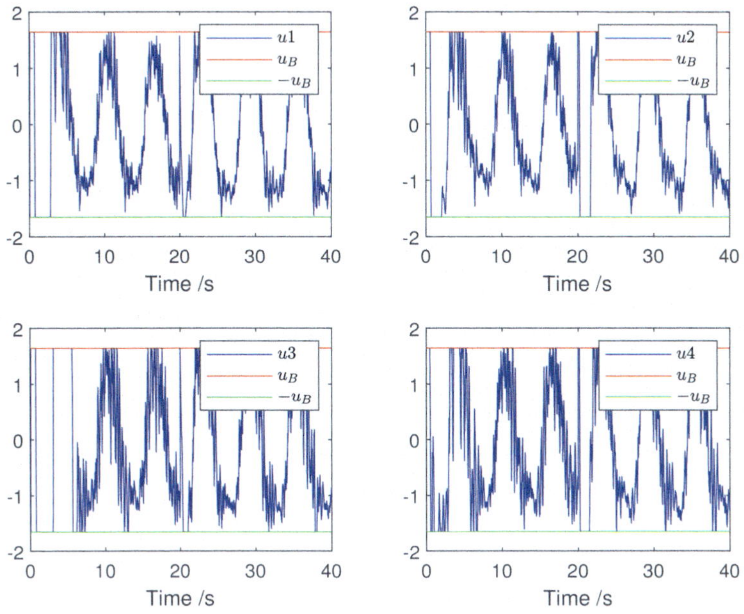

Fig. 8.10 Controller outputs $u_i(i = 1, 2, 3, 4)$ (case 2)

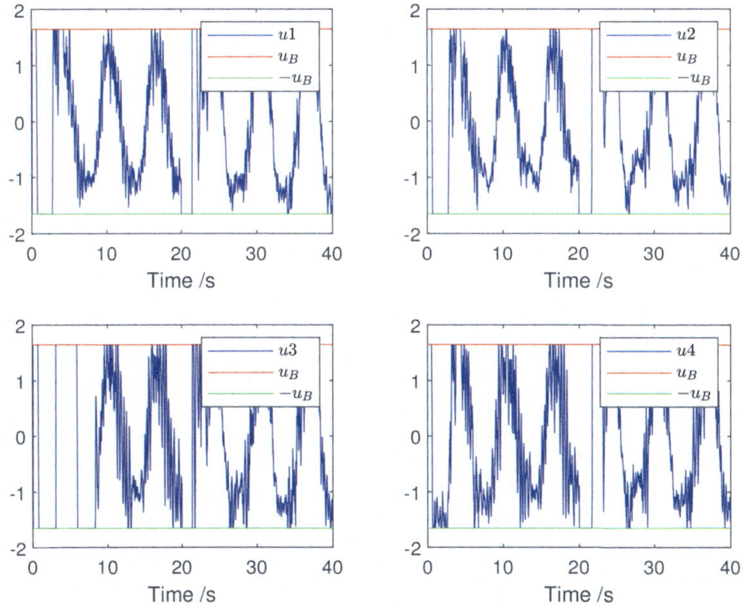

Fig. 8.11 Controller outputs $u_i(i = 1, 2, 3, 4)$ (case 3)

References

Cao L, Li H, Dong G, Lu R (2021) Event-triggered control for multiagent systems with sensor faults and input saturation. IEEE Trans Syst Man Cybernet Syst 51(6):3855–3866

Chen M, Tao G, Jiang B (2015) Dynamic surface control using neural networks for a class of uncertain nonlinear systems with input saturation. IEEE Trans Neural Netw Learn Syst 26(9):2086–2097

Chen C, Lewis FL, Xie S, Modares H, Liu Z, Zuo S, Davoudi A (2019) Resilient adaptive and H_∞ controls of multi-agent systems under sensor and actuator faults. Automatica 102:19–26

Davoodi M, Meskin N, Khorasani K (2016) Simultaneous fault detection and consensus control design for a network of multi-agent systems. Automatica 66:185–194

Deng C, Che WW (2021) Fault-tolerant fuzzy formation control for a class of nonlinear multiagent systems under directed and switching topology. IEEE Trans Syst Man Cybernet Syst 51(9):5456–5465

Dong G, Li H, Ma H, Lu R (2021) Finite-time consensus tracking neural network FTC of multi-agent systems. IEEE Trans Neural Netw Learn Syst 32(2):653–662

Du Y, Wang Y, Zuo Z, Zhang W (2020) Containment control for distributed networks subject to multiplicative and additive noises with stochastic approximation-type protocols. Int J Robust Nonlinear Control 30(2):665–684

Du H, Wen G, Wu D, Cheng Y, Lü J (2020) Distributed fixed-time consensus for nonlinear heterogeneous multi-agent systems. Automatica 113:108797

Hu J, Feng G (2010) Distributed tracking control of leader-follower multi-agent systems under noisy measurement. Automatica 46(8):1382–1387

Hua C, Li Y, Guan X (2017) Leader following consensus for high-order nonlinear stochastic multiagent systems. IEEE Trans Cybernet 47(8):1882–1891

Kommuri SK, Lee SB, Veluvolu KC (2018) Robust sensors-fault-tolerance with sliding mode estimation and control for PMSM drives. IEEE/ASME Trans Mechatron 23(1):17–28

Li K, Mu X (2019) Containment control of stochastic multiagent systems with semi-Markovian switching topologies. Int J Robust Nonlinear Control 29(14):4943–4955

Li T, Zhang JF (2008) Asymptotically optimal decentralized control for large population stochastic multiagent systems. IEEE Trans Autom Control 53(7):1643–1660

Li Z, Duan Z, Lewis FL (2014) Distributed robust consensus control of multi-agent systems with heterogeneous matching uncertainties. Automatica 50(3):883–889

Li Y, Yang T, Tong S (2020) Adaptive neural networks finite-time optimal control for a class of nonlinear systems. IEEE Trans Neural Netw Learn Syst 31(11):4451–4460

Li Y, Qu F, Tong S (2021) Observer-based fuzzy adaptive finite-time containment control of nonlinear multiagent systems with input delay. IEEE Trans Cybernet 51(1):126–137

Li H, Wu Y, Chen M (2021) Adaptive fault-tolerant tracking control for discrete-time multiagent systems via reinforcement learning algorithm. IEEE Trans Cybernet 51(3):1163–1174

Liang H, Guo X, Pan Y, Huang T (2021) Event-triggered fuzzy bipartite tracking control for network systems based on distributed reduced-order observers. IEEE Trans Fuzzy Syst 29(6):1601–1614

Liang H, Zhang L, Sun Y, Huang T (2021) Containment control of semi-Markovian multiagent systems with switching topologies. IEEE Trans Syst Man Cybernet Syst 51(6):3889–3899

Liang H, Liu G, Zhang H, Huang T (2021) Neural-network-based event-triggered adaptive control of nonaffine nonlinear multiagent systems with dynamic uncertainties. IEEE Trans Neural Netw Learn Syst 32(5):2239–2250

Lin G, Li H, Ma H, Yao D, Lu R (2022) Human-in-the-loop consensus control for nonlinear multi-agent systems with actuator faults. IEEE/CAA J Automatica Sinica 9(1):111–122

Long J, Wang W, Huang J, Fan H (2020) Distributed adaptive consensus tracking for uncertain high-order nonlinear multiagent systems with event-triggered communication. Opt Control Appl Methods 41(6):2077–2093

Ma H, Wang Z, Wang D, Liu D, Yan P, Wei Q (2016) Neural-network-based distributed adaptive robust control for a class of nonlinear multiagent systems with time delays and external noises. IEEE Trans Syst Man Cybernet Syst 46(6):750–758

Ma H, Ren H, Zhou Q, Lu R, Li H (2022) Approximation-based Nussbaum gain adaptive control of nonlinear systems with periodic disturbances. IEEE Trans Syst Man Cybernet Syst 52(4):2591–2600

Ma H, Li H, Lu R, Huang T (2020) Adaptive event-triggered control for a class of nonlinear systems with periodic disturbances. Sci China Inf Sci 63(5):150212

Pan Y, Du P, Xue H, Lam HK (2021) Singularity-free fixed-time fuzzy control for robotic systems with user-defined performance. IEEE Trans Fuzzy Syst 29(8):2388–2398

Pan Y, Wu Y, Lam HK (2022) Security-based fuzzy control for nonlinear networked control systems with dos attacks via a resilient event-triggered scheme. IEEE Trans Fuzzy Syst 30(10):4359–4368

Pan Y, Li Q, Liang H, Lam HK (2022) A novel mixed control approach for fuzzy systems via membership functions online learning policy. IEEE Trans Fuzzy Syst 30(9):3812–3822

Qin J, Li M, Shi Y, Ma Q, Zheng WX (2019) Optimal synchronization control of multiagent systems with input saturation via off-policy reinforcement learning. IEEE Trans Neural Netw Learn Syst 30(1):85–96

Ren H, Karimi HR, Lu R, Wu Y (2020) Synchronization of network systems via aperiodic sampled-data control with constant delay and application to unmanned ground vehicles. IEEE Trans Indus Electron 67(6):4980–4990

Ren H, Lu R, Xiong J, Wu Y, Shi P (2020) Optimal filtered and smoothed estimators for discrete-time linear systems with multiple packet dropouts under Markovian communication constraints. IEEE Trans Cybernet 50(9):4169–4181

Ren H, Lu R, Xiong J, Xu Y (2020) Optimal estimation for discrete-time linear system with communication constraints and measurement quantization. IEEE Trans Syst Man Cybernet Syst 50(5):1932–1942

Rezaee H, Abdollahi F (2017) Consensus problem in high-order multiagent systems with Lipschitz nonlinearities and jointly connected topologies. IEEE Trans Syst Man Cybernet Syst 47(5):741–748

Sakthivel R, Sakthivel R, Kaviarasan B, Alzahrani F (2018) Leader-following exponential consensus of input saturated stochastic multi-agent systems with Markov jump parameters. Neurocomputing 287:84–92

Su H, Chen MZ, Lam J, Lin Z (2013) Semi-global leader-following consensus of linear multi-agent systems with input saturation via low gain feedback. IEEE Trans Circuits Syst I: Regular Pap 60(7):1881–1889

Su H, Zhang J, Zeng Z (2020) Formation-containment control of multi-robot systems under a stochastic sampling mechanism. Sci China Technol Sci 63(6):1025–1034

Tong S, Min X, Li Y (2020) Observer-based adaptive fuzzy tracking control for strict-feedback nonlinear systems with unknown control gain functions. IEEE Trans Cybernet 50(9):3903–3913

Tong S, Li Y, Li Y, Liu Y (2011) Observer-based adaptive fuzzy backstepping control for a class of stochastic nonlinear strict-feedback systems. IEEE Trans Syst Man Cybernet Part B (Cybernetics) 41(6):1693–1704

Wang A, Mu B, Shi Y (2017) Consensus control for a multi-agent system with integral-type event-triggering condition and asynchronous periodic detection. IEEE Trans Indus Electron 64(7):5629–5639

Wang X, Su H, Chen MZ, Wang X (2018) Observer-based robust coordinated control of multiagent systems with input saturation. IEEE Trans Neural Netw Learn Syst 29(5):1933–1946

Wen G, Wang P, Huang T, Yu W, Sun J (2017) Robust neuro-adaptive containment of multileader multiagent systems with uncertain dynamics. IEEE Trans Syst Man Cybernet Syst 49(2):406–417

Wu Y, Pan Y, Chen M, Li H (2021) Quantized adaptive finite-time bipartite NN tracking control for stochastic multiagent systems. IEEE Trans Cybernet 51(6):2870–2881

Xiong S, Hou Z (2022) Data-driven formation control for unknown MIMO nonlinear discrete-time multi-agent systems with sensor fault. IEEE Trans Neural Netw Learn Syst 33(12):7728–7742

Yao D, Li H, Lu R, Shi Y (2020) Distributed sliding-mode tracking control of second-order nonlinear multiagent systems: an event-triggered approach. IEEE Trans Cybernet 50(9):3892–3902
Zhang D, Shi P, Yu L (2018) Containment control of linear multiagent systems with aperiodic sampling and measurement size reduction. IEEE Trans Neural Netw Learn Syst 29(10):5020–5029
Zhou Q, Zhao S, Li H, Lu R, Wu C (2019) Adaptive neural network tracking control for robotic manipulators with dead zone. IEEE Trans Neural Netw Learn Syst 30(12):3611–3620
Zhou Q, Wang W, Ma H, Li H (2021) Event-triggered fuzzy adaptive containment control for nonlinear multi-agent systems with unknown Bouc-Wen hysteresis input. IEEE Trans Fuzzy Syst 29(4):731–741
Zhu JW, Yang GH, Zhang WA, Yu L (2018) Cooperative fault tolerant tracking control for multiagent systems: an intermediate estimator-based approach. IEEE Trans Cybernet 48(10):2972–2980

Chapter 9
Aperiodic Sampled Data Control with Constant Delay

Abstract In this chapter, the synchronization issue for network systems with nonlinear dynamics is considered. Together with Zero-Order Holder (ZOH), the aperiodic sampled data control law is utilized. Compared with the traditional periodic sampled data control method, this approach demonstrates more greater flexibility. Adopting input delay approach, the initial sampled data system is remodeled by continuous time system involving time-varying delays in the control signals. For the purpose of designing the sampling controllers suffering constant delays, an updated Lyapunov functional is developed from the augmentation of Wirtinger's inequality. Such a Lyapunov functional results in efficient and simplified synchronization conditions. A sufficient condition for synchronizability of network systems is set up. Then, for the case of unstable systems with some constant delays, a fresh discretized Lyapunov functional is introduced. At last, we utilize the numerical simulation outcomes to prove the efficacy and advantage of our algorithm. Moreover, based on the network of unmanned ground vehicle systems, the experiment results in a real scenario are provided to illustrate the effectiveness of the designed synchronization scheme.

9.1 Introduction

Collaboration in network systems has exerted a tremendous fascination on individuals in the last few decades (Karimi et al. 2010; Lu et al. 2016; Shi and Yu 2009, 2011; Zhang et al. 2018; Liang et al. 2020; Shen et al. 2018; Mu and Shi 2018; Wu et al. 2018; Li et al. 2013; Ma et al. 2019). A multitude of issues can be addressed via cooperative control such as the rendezvous problem, formation control, swarm analysis and flocking, and synchronization seeking (Semsar-Kazerooni and Khorasani 2010). Synchronization seeking or synchronous control was intended to design an appropriate network control protocol or scheme for every system on the basis of local state messages of its adjacent systems (Chen et al. 2014). Ultimately, such a scheme gives rise to the consequence that every individual system of the network systems reaches a shared isochronous state. A host of efforts and achievements have been made for this difficulty over recent years. Consequently, various kinds of synchronous control method for network systems are designed successively. In Sun et al.

H. Ren et al., *Intelligent Control of Nonlinear Systems and Nonlinear Multi-Agent Systems with Complex Constraints*, Intelligent Control and Learning Systems 16,
https://doi.org/10.1007/978-981-96-0320-6_9

(2015), together with a synchronization approach, a neoteric coordinated distributed controller was proposed under network systems structure. Distributed generators were applied to enhance the utilization of energy between main grid and Energy Internet. Robust synchronization of uncertain linear network systems was studied in Trentelman et al. (2013). Indetermination in the shape of additive destabilizations of the transfer matrices of the nominal kinetics was allowed. The synchronization issue for network systems with switching and directed interconnected topologies was addressed in Thunberg et al. (2014). In addition, a multitude of complex network systems (Deng et al. 2019, 2017a, b), e.g., high-order dynamics (Li et al. 2011; Sun et al. 2020), impulsive effect (Xiong et al. 2012), and stochastic disturbance (Lin et al. 2008) have been studied in literatures.

With regard to network systems, much attention has been poured into synchronization issue with continuous control methods. However, attributable to some immanent characteristics of network environment, sometimes continuous control law is not suitable. To solve such problem, compared with continuous one, sampled data control strategy is a better selection. The control inputs are updated in a discrete way, while the operation status of sampled data systems is continuous (Shi 1998; Gao et al. 2009; Meng and Chen 2012; Liu et al. 2012, 2013; Guo and Dimarogonas 2013). Though periodic sampling method is broadly applied in some scenarios, for the sake of further reducing energy consumption, it is essential to utilize aperiodic sampling approach. For the purpose of analyzing acyclic sampled data systems, some pivotal issues were investigated in literature. In Oishi and Fujioka (2010), taking networked and embedded systems as the background of application, stability analysis for non-periodic sampled data system was studied. In Wu et al. (2016), utilizing acyclic sampled data control strategy, the synchronization problem was investigated for network systems with nonlinear dynamics. In Suh (2008), authors rebuilt the continuous system with time-variant sampling time lags as a discrete-time variant system. In Fridman (2010), based on input delay method, the original sampling system was rewritten as a time delay system.

On account of the actuality that signals and data packets are transmitted over a wireless network under most circumstances, time delay usually arises (Liu and Fridman 2012; Zhang and Tian 2010; Ali and Vadivel 2018; Ali et al. 2018a, b; Zhang et al. 2019; Xia et al. 2021). The issues of time delay are broadly investigated for network systems. In Yan and Huang (2017), authors addressed the problems of cooperative output regulation for jointly connected switching network systems with time delay. The distributed control strategy consists of a purely decentralized control law and a distributed switched observer. In Ma et al. (2017), introducing the smart leader, the synchronization issues were considered for delayed network systems with continuous single-integrator dynamics. In Lin and Jia (2010), the synchronization problems were concerned for a class of second-order continuous network systems with time delay and jointly-connected topologies. Furthermore, a neighbor-based linear strategy was utilized. In literature, the input delay approach is a popular method to transfer the sampling control signals into continuous functions. Adopting this method, researchers can express the closed-loop systems as switching systems with multiple sequential time-variant delays (Liu et al. 2012; Liu and Fridman 2014).

In the existing literature, many achievements and progress on synchronization issues have been made for network systems. Nevertheless, there are still many important problems to be solved urgently. In this chapter, we summarize the contributions and novelty as follows.

1. To reduce the communication load among UGVs, sampling control law is utilized, whose effectiveness is demonstrated by experiments.
2. Derived from the augmented Wirtinger's inequality, a novel Lyapunov functional is established for stable systems, which leads more greater flexibility than (Wu et al. 2016).
3. For the situation of unstable systems with some constant delays, a discretized Lyapunov functional is introduced.

9.1.1 Notations

Throughout the chapter, $\mathbb{R}^n$ represents n-dimensional real Euclidean space. The notation $\|\cdot\|$ represents the norm of vector or matrix. For a symmetric matrix, the $*$ represents the symmetric term. $P \otimes Q$ stands for the Kronecker product.

9.2 Problem Formulation and Preliminaries

We utilize $\mathfrak{G} = (\mathfrak{V}, \mathfrak{E}, \mathfrak{A})$ to express a digraph or an oriented graph with vertices $\mathfrak{V} = \{v_1, v_2, \ldots, v_N\}$, directed edges $\mathfrak{E} \subseteq \mathfrak{V} \times \mathfrak{V}$, and the adjacent matrix $\mathfrak{A} = [a_{ij}]_{N\times N}$. We represent $\mathfrak{G} = (\mathfrak{V}, \mathfrak{E}, \mathfrak{A})$ by $\mathfrak{G}(\mathfrak{A})$ briefly. If vertex l can send information to vertex s, then $\mathfrak{G}(\mathfrak{A})$ contains an directed edge $(v_l, v_s) \in \mathfrak{E}$. In adjacent matrix $\mathfrak{A}$, $a_{ls} = 1$ means that $(v_l, v_s) \in \mathfrak{G}(\mathfrak{A})$ and $a_{ls} = 0$ denotes $(v_l, v_s) \notin \mathfrak{G}(\mathfrak{A})$.

With regard to the digraph $\mathfrak{G}(\mathfrak{A})$, its Laplacian matrix $\mathfrak{L} = [m_{ls}]_{N\times N}$ is

$$\begin{cases} m_{ls} = -a_{ls}, & l \neq s \\ m_{ll} = -\sum_{k=1,k\neq l}^{N} m_{lk}, & l = s \end{cases}. \tag{9.1}$$

Based on the diffusion property, $\mathfrak{L}$ satisfies $\sum_{s=1}^{N} m_{ls} = 0$ for a strong digraph $\mathfrak{G}(\mathfrak{A})$.

Consider a class of network systems with N nonlinear nodes or subsystems

$$\dot{x}_i(t) = Ax_i(t) + f(x_i(t)) + Bu_i(t). \tag{9.2}$$

For the ith node, $x_i(t) = [x_{i1}(t), \ldots, x_{in}(t)]^T \in \mathbb{R}^n$ denotes the state. $u_i(t) \in \mathbb{R}^p$ stands for the designed control input. A and B are constant matrices with appropriate dimensions. The nonlinear vector-valued function $f(x_i(t)) = [f(x_{i1}(t)), \ldots, f(x_{in}(t))]^T \in \mathbb{R}^n$ signifies the inherent nonlinear dynamics of system i, which is supposed to meet the Lipschitz condition as follows.

Assumption 9.1 Every nonlinear function $g_\kappa(\cdot)$ in (9.2) is Lipschitz continuous, if there exist positive parameters d_κ so that

$$\|g_\kappa(\rho_1) - g_\kappa(\rho_2)\| \leq d_\kappa \|\rho_1 - \rho_2\|, \quad \kappa = 1, \ldots, n \tag{9.3}$$

hold for any $\rho_1, \rho_2 \in \mathbb{R}$.

Remark 9.1 When the derivative of nonlinear function $g(\cdot)$ is bounded, $g(\cdot)$ matches the Lipschitz condition. Therefore, in this chapter, Assumption 9.1 is utilized to depict the nonlinear characteristics for the network UGV systems experimentally. A detailed description on UGVs used for experiments can be found in the section of simulations.

For network nonlinear systems (9.2), the goal of this chapter is to design a sampled data synchronous controller u_l for each subsystem l. The definition and lemmas are recalled as follows before giving key results.

Definition 9.1 (Ren and Beard 2005) If the formula holds for any $l, s = 1, \ldots, N$,

$$\lim_{t\to\infty} \|x_l(t) - x_s(t)\| = 0 \tag{9.4}$$

the network nonlinear systems (9.2) are regarded as achieving the synchronization.

Lemma 9.1 (Ma et al. 2014) *For a strong digraph $\mathfrak{G}$, its Laplacian matrix $\mathfrak{L}$ meets $\sum_{s=1}^N m_{ls} = 0$. Furthermore, $\mathfrak{L}$ has eigenvalue 0. The normalized left eigenvector $\varsigma = [\varsigma_1, \varsigma_2, \ldots, \varsigma_N]^T \in \mathbb{R}^N$ meeting $\sum_{i=1}^N \varsigma_i = 1$, i.e., $\varsigma^T \cdot \mathfrak{L} = 0$.*

Lemma 9.2 (Lu and Ho 2010) *For a strong digraph $\mathfrak{G}$, let $\varsigma = [\varsigma_1, \varsigma_2, \ldots, \varsigma_N]^T$ be the left normalized eigenvector of Laplacian matrix $\mathfrak{L}$ with eigenvalue 0, $\Xi = diag\{\varsigma_1, \varsigma_2, \ldots, \varsigma_N\} > 0$, and $W_{N\times N} = \Xi - \varsigma\varsigma^T$. Accordingly, $\forall \mathfrak{P} > 0$, the following formulas hold:*

$$\begin{aligned} & x^T(t)(W\mathfrak{L} \otimes \mathfrak{P}C)g(x(t)) \\ =& -\frac{1}{2}\sum_{l=1}^{N}\sum_{s=1,s\neq l}^{N} \varsigma_l m_{ls}\big(x_l(t) - x_s(t)\big)^T \mathfrak{P}C\big[g(x_l(t)) - g(x_s(t))\big] \end{aligned} \tag{9.5}$$

$$\begin{aligned} & x^T(t)(W \otimes \mathfrak{P}C)g(x(t)) \\ =& \frac{1}{2}\sum_{l=1}^{N}\sum_{s=1,s\neq l}^{N} \varsigma_l \varsigma_s\big(x_l(t) - x_s(t)\big)^T \mathfrak{P}C\big[g(x_l(t)) - g(x_s(t))\big]. \end{aligned} \tag{9.6}$$

Derived from the augmented Wirtinger's inequality, a novel Lyapunov functional will be introduced.

Lemma 9.3 (Liu and Fridman 2012) *Suppose that $\zeta(t)$ has square integrable first-order derivative in $[u, v)$ and $\zeta(u) = 0$. Then for any matrix $\mathfrak{P} > 0$ with suitable dimensions, we have*

$$\frac{\pi^2}{4(v-u)^2}\int_u^v \zeta^T(\xi)\mathfrak{P}\zeta(\xi)d\xi \le \int_u^v \dot{\zeta}^T(\xi)\mathfrak{P}\dot{\zeta}(\xi)d\xi. \tag{9.7}$$

9.3 Sampled Data Synchronous Controller

For subsystem $l = 1, 2, \ldots, N$, we adopt the following synchronous control law

$$u_l(t) = \psi K \sum_{s=1, s\neq l}^{N} a_{ls}\left[x_s(t_k - \varpi) - x_l(t_k - \varpi)\right] \tag{9.8}$$

where $\psi > 0$ stands for the coupling strength parameter, and $t \in [t_k, t_{k+1}),\ k \in \mathbb{N}$. K represents the gain matrix of controller. t_k denotes the instant when the ZOH updates. Furthermore, ϖ is the constant delay. Applying ZOH, the control input holds from t_k to t_{k+1}. It can be obtained that $\varpi = t_0 < t_1 < \cdots < t_k < \cdots < \lim_{k\to\infty} t_k = +\infty$. Suppose that the sampled data time lags are alterable below the upper bound b_u as

$$(t_{k+1} - \varpi) - (t_k - \varpi) = t_{k+1} - t_k \le b_u, \quad k \in \mathbb{N}. \tag{9.9}$$

By defining $\kappa_M \triangleq b_u + \varpi$, one has

$$t_{k+1} - (t_k - \varpi) = t_{k+1} - t_k + \varpi \le \kappa_M, \quad k \in \mathbb{N} \tag{9.10}$$

where κ_M stands for the maximum time interval from $t_k - \varpi$ to t_{k+1}.

Define

$$\begin{aligned} x(t) &= \left[x_1^T(t)\ \ldots\ x_N^T(t)\right]^T \\ x(t_k - \varpi) &= \left[x_1^T(t_k - \varpi)\ \ldots\ x_N^T(t_k - \varpi)\right]^T \\ F(x(t)) &= \left[f^T(x_1(t))\ \ldots\ f^T(x_N(t))\right]^T. \end{aligned}$$

Substituting (9.8) into (9.2) leads to

$$\dot{x}(t) = -\psi(\mathfrak{L} \otimes BK)x(t_k - \varpi) + (I_N \otimes A)x(t) + F(x(t)) \tag{9.11}$$

where $t \in [t_k, t_{k+1})$, $k \in \mathbb{N}$. Inspired by Fridman (2010), Liu et al. (2012), Liu and Fridman (2014), we adopt input delay method. By setting $\kappa(t) = t - t_k + \varpi$, $t \in [t_k, t_{k+1})$, from (9.10) we obtain $\varpi \le \kappa(t) < t_{k+1} - t_k + \varpi \le \kappa_M$ and $\dot{\kappa}(t) = 1$ for $t \neq t_k$. Then, (9.11) is rewritten by

$$\dot{x}(t) = -\psi(\mathfrak{L} \otimes BK)x(t - \kappa(t)) + (I_N \otimes A)x(t) + F(x(t)). \tag{9.12}$$

9.3.1 Synchronization by Simple Lyapunov Functional

For the synchronization issue of (9.2), the first sufficient condition is given via Theorem 9.1.

Theorem 9.1 *For a strong digraph $\mathfrak{G}$, given constant delay $\varpi > 0$, gain matrix K, and sampling upper bound $b_u > 0$, if exist some matrices $Z_1 > 0$, $Z_2 > 0$, $Q > 0$, $P > 0$, $R > 0$, $J > 0$, diagonal matrix $G > 0$, and $\mathcal{N}_s$, $(s = 1, \ldots, 6)$, so that the following inequality:*

$$\mathfrak{M}^{ij} = \begin{bmatrix} \mathfrak{M}_{11} & \mathfrak{M}_{12} & \mathfrak{M}^{ij}_{13} & \mathfrak{M}_{14} & \mathfrak{M}_{15} & \mathfrak{M}_{16} \\ * & \mathfrak{M}_{22} & \mathfrak{M}^{ij}_{23} & 0 & 0 & \mathfrak{M}_{26} \\ * & * & \mathfrak{M}^{ij}_{33} & \mathfrak{M}^{ij}_{34} & \mathfrak{M}^{ij}_{35} & \mathfrak{M}^{ij}_{36} \\ * & * & * & \mathfrak{M}_{44} & 0 & \mathfrak{M}_{46} \\ * & * & * & * & \mathfrak{M}_{55} & \mathfrak{M}_{56} \\ * & * & * & * & * & \mathfrak{M}_{66} \end{bmatrix} < 0 \tag{9.13}$$

holds for any $i, j = 1, \ldots, N$, where $\bar{\psi}_{ij} = \psi(l_{ij}/\varsigma_j)$ and

$$\begin{aligned}
\mathfrak{M}_{11} =& Q - R + \mathcal{N}_1 A + A^T \mathcal{N}_1^T + D^T G D + \frac{b_u^2}{\varpi \kappa_M} Z_2 \\
\mathfrak{M}_{12} =& P - \mathcal{N}_1 + A^T \mathcal{N}_2^T \\
\mathfrak{M}^{ij}_{13} =& \bar{\psi}_{ij} \mathcal{N}_1 B K + A^T \mathcal{N}_3^T \\
\mathfrak{M}_{14} =& R + A^T \mathcal{N}_4^T - \frac{b_u}{\varpi} Z_2 \\
\mathfrak{M}_{15} =& A^T \mathcal{N}_5^T + \frac{b_u}{\kappa_M} Z_2 \\
\mathfrak{M}_{16} =& \mathcal{N}_1 + A^T \mathcal{N}_6^T \\
\mathfrak{M}_{22} =& b_u^2 (J + Z_2) + \varpi^2 R - \mathcal{N}_2 - \mathcal{N}_2^T \\
\mathfrak{M}^{ij}_{23} =& \bar{\psi}_{ij} \mathcal{N}_2 B K - \mathcal{N}_3 \\
\mathfrak{M}_{26} =& \mathcal{N}_2 - \mathcal{N}_6^T \\
\mathfrak{M}^{ij}_{33} =& \bar{\psi}_{ij} (\mathcal{N}_3 B K + K^T B^T \mathcal{N}_3^T) - \frac{\pi^2}{4} J \\
\mathfrak{M}^{ij}_{34} =& \bar{\psi}_{ij} K^T B^T \mathcal{N}_4^T + \frac{\pi^2}{4} J \\
\mathfrak{M}^{ij}_{35} =& \bar{\psi}_{ij} K^T B^T \mathcal{N}_5^T \\
\mathfrak{M}^{ij}_{36} =& \bar{\psi}_{ij} K^T B^T \mathcal{N}_6^T + \mathcal{N}_3 \\
\mathfrak{M}_{44} =& Z_1 - Q - R + \frac{b_u}{\varpi} Z_2 - \frac{\pi^2}{4} J \\
\mathfrak{M}_{46} =& \mathcal{N}_4
\end{aligned}$$

$$\mathfrak{M}_{55} = - Z_1 - \frac{b_u}{\kappa_M} Z_2$$
$$\mathfrak{M}_{56} = \mathcal{N}_5$$
$$\mathfrak{M}_{66} = \mathcal{N}_6 + \mathcal{N}_6^T - G$$

the network nonlinear systems (9.2) *has the ability to achieve synchronization.*

Proof First of all, with regard to system (9.11), consider the Lyapunov functional:

$$V\big(x(t)\big) = V_J\big(x(t)\big) + V_P\big(x(t)\big) + V_Z\big(x(t)\big) \tag{9.14}$$

where

$$\begin{aligned} V_J\big(x(t)\big) =& (\kappa_M - \varpi)^2 \int_{t_k-\varpi}^{t} \dot{x}^T(\gamma)(W \otimes J)\dot{x}(\gamma)d\gamma - \frac{\pi^2}{4} \int_{t_k-\varpi}^{t-\varpi} [x(t_k - \varpi) \\ & - x(\gamma)]^T (W \otimes J)[x(t_k - \varpi) - x(\gamma)]d\gamma \\ V_P\big(x(t)\big) =& x^T(t)(W \otimes P)x(t) + \int_{t-\varpi}^{t} x^T(\gamma)(W \otimes Q)x(\gamma)d\gamma \\ & + \varpi \int_{-\varpi}^{0} \int_{t+\theta}^{t} \dot{x}^T(\gamma)(W \otimes R)\dot{x}(\gamma)d\gamma d\theta \\ V_Z\big(x(t)\big) =& \int_{t-\kappa_M}^{t-\varpi} x^T(\gamma)(W \otimes Z_1)x(\gamma)d\gamma \\ & + (\kappa_M - \varpi) \int_{-\kappa_M}^{-\varpi} \int_{t+\theta}^{t} \dot{x}^T(\gamma)(W \otimes Z_2)\dot{x}(\gamma)d\gamma d\theta. \end{aligned}$$

Remark 9.2 The Lyapunov functionals V_J and V_P were adopted in Liu and Fridman (2012). Sometimes, more complicated linear matrix inequality (LMI) conditions by Jensen's inequality and time-independent Lyapunov functionals can show a lower restrictiveness than Wirtinger-based conditions. Therefore, an updated functional $V_J + V_P + V_Z$ is adopted for the purpose of obtaining less conservative conditions. □

Calculating the time derivative of functional (9.14) results in

$$\begin{aligned} \dot{V}_J\big(x(t)\big) =& b_u^2 \dot{x}^T(t)(W \otimes J)\dot{x}(t) - \frac{\pi^2}{4}[x(t - \varpi) - x(t_k - \varpi)]^T \\ & \times (W \otimes J)[x(t - \varpi) - x(t_k - \varpi)] \\ \dot{V}_P\big(x(t)\big) =& 2x^T(t)(W \otimes P)\dot{x}(t) + x^T(t)(W \otimes Q)x(t) \end{aligned}$$

$$
\begin{aligned}
&- x^T(t-\varpi)(W\otimes Q)x(t-\varpi) + \varpi^2\dot{x}^T(t)(W\otimes R)\dot{x}(t)\\
&- \varpi\int_{t-\varpi}^{t}\dot{x}^T(\gamma)(W\otimes R)\dot{x}(\gamma)d\gamma\\
\dot{V}_Z\big(x(t)\big) = {} & x^T(t-\varpi)(W\otimes Z_1)x(t-\varpi) - x^T(t-\kappa_M)\\
&\times (W\otimes Z_1)x(t-\kappa_M) + b_u^2\dot{x}^T(t)(W\otimes Z_2)\dot{x}(t)\\
&- b_u\int_{t-\kappa_M}^{t-\varpi}\dot{x}^T(\gamma)(W\otimes Z_2)\dot{x}(\gamma)d\gamma.
\end{aligned}
$$

By taking advantage of Jensen's inequality, we obtain

$$
\begin{aligned}
&- \varpi\int_{t-\varpi}^{t}\dot{x}^T(\gamma)(W\otimes R)\dot{x}(\gamma)d\gamma\\
&\le -\int_{t-\varpi}^{t}\dot{x}^T(\gamma)d\gamma(W\otimes R)\int_{t-\varpi}^{t}\dot{x}(\gamma)d\gamma\\
&= -[x(t)-x(t-\varpi)]^T(W\otimes R)[x(t)-x(t-\varpi)]
\end{aligned}
$$

and similarly

$$
\begin{aligned}
&- b_u\int_{t-\kappa_M}^{t-\varpi}\dot{x}^T(\gamma)(W\otimes Z_2)\dot{x}(\gamma)d\gamma\\
&\le \frac{b_u}{\varpi}[x(t)-x(t-\varpi)]^T(W\otimes Z_2)[x(t)-x(t-\varpi)]\\
&- \frac{b_u}{\kappa_M}[x(t)-x(t-\kappa_M)]^T(W\otimes Z_2)[x(t)-x(t-\kappa_M)].
\end{aligned}
$$

Let

$$
\xi(t) \triangleq \begin{bmatrix} x(t)\\ \dot{x}(t)\\ x(t_k-\varpi)\\ x(t-\varpi)\\ x(t-\kappa_M)\\ F(x(t)) \end{bmatrix},\quad \mathcal{N} \triangleq \begin{bmatrix} \mathcal{N}_1\\ \mathcal{N}_2\\ \mathcal{N}_3\\ \mathcal{N}_4\\ \mathcal{N}_5\\ \mathcal{N}_6 \end{bmatrix}.
$$

Then, we can see that

$$2\xi^T(t)(W \otimes \mathcal{N})\big[F(x(t)) - \psi(\mathfrak{L} \otimes BK)x(t_k - \varpi) + (I_N \otimes A)x(t) - \dot{x}(t)\big] = 0. \tag{9.15}$$

Denote $\widetilde{\varphi} = \varphi_i - \varphi_j$. Thus based on (9.3) of Assumption 9.1, we can find a matrix $D = \text{diag}\{d_1, d_2, \ldots, d_n\}$ such that

$$-\widetilde{f}^T(x(t))G\widetilde{f}(x(t)) + \widetilde{x}^T(t)D^T GD\widetilde{x}(t) \geq 0. \tag{9.16}$$

From (9.15), (9.16), and Lemma 9.2, it can be seen that

$$\begin{aligned}
\dot{V}_J\big(x(t)\big) = \frac{1}{2}\sum_{l=1}^{N}\sum_{s=1,s\neq l}^{N} \varsigma_l\varsigma_s\Big\{& b_u^2\widetilde{\dot{x}}^T(t)J\widetilde{\dot{x}}(t) - \frac{\pi^2}{4}[\widetilde{x}(t-\varpi) \\
&- \widetilde{x}(t_k - \varpi)]^T J[\widetilde{x}(t-\varpi) - \widetilde{x}(t_k - \varpi)]\Big\} \\
\dot{V}_P\big(x(t)\big) \leq \frac{1}{2}\sum_{l=1}^{N}\sum_{s=1,s\neq l}^{N} \varsigma_l\varsigma_s\Big\{& 2\widetilde{x}^T(t)P\widetilde{\dot{x}}(t) + \widetilde{x}^T(t)Q\widetilde{x}(t) \\
&- \widetilde{x}^T(t-\varpi)Q\widetilde{x}(t-\varpi) + \varpi^2\widetilde{\dot{x}}^T(t)R\widetilde{\dot{x}}(t) \\
&- [\widetilde{x}(t) - \widetilde{x}(t-\varpi)]^T R[\widetilde{x}(t) - \widetilde{x}(t-\varpi)]\Big\} \\
\dot{V}_Z\big(x(t)\big) \leq \frac{1}{2}\sum_{l=1}^{N}\sum_{s=1,s\neq l}^{N} \varsigma_l\varsigma_s\Big\{& \widetilde{x}^T(t-\varpi)Z_1\widetilde{x}(t-\varpi) \\
&- \widetilde{x}^T(t-\kappa_M)Z_1\widetilde{x}(t-\kappa_M) + b_u^2\widetilde{\dot{x}}^T(t)Z_2\widetilde{\dot{x}}(t) \\
&+ \frac{b_u}{\varpi}[\widetilde{x}(t) - \widetilde{x}(t-\varpi)]^T Z_2[\widetilde{x}(t) - \widetilde{x}(t-\varpi)] \\
&- \frac{b_u}{\kappa_M}[\widetilde{x}(t) - \widetilde{x}(t-\kappa_M)]^T Z_2[\widetilde{x}(t) - \widetilde{x}(t-\kappa_M)]\Big\}
\end{aligned}$$

which leads to

$$\dot{V}\big(x(t)\big) \leq \frac{1}{2}\sum_{l=1}^{N}\sum_{s=1,s\neq l}^{N} \varsigma_l\varsigma_s\widetilde{\xi}^T(t)\mathfrak{M}^{ij}\widetilde{\xi}(t). \tag{9.17}$$

From (9.13), it can be obtained that $V\big(x(t)\big) \leq V\big(x(0)\big)$ and $\dot{V}\big(x(t)\big) < 0$, which reveals the boundedness of $V\big(x(t)\big)$. Accordingly, $x^T(t)(W \otimes P)x(t)$ is also bounded and

$$\begin{aligned}
\varsigma_l\varsigma_s\lambda_{min}(P)\|\widetilde{x}(t)\|^2 &\leq \frac{1}{2}\sum_{l=1}^{N}\sum_{s=1,s\neq l}^{N} \varsigma_l\varsigma_s\widetilde{x}^T(t)P\widetilde{x}(t) \\
&= \widetilde{x}^T(t)(W \otimes P)\widetilde{x}(t) = O(e^{-\varsigma t}).
\end{aligned} \tag{9.18}$$

Therefore, based on Definition 9.1, we can obtain that the synchronization in network systems (9.2) can be guaranteed. □

Next, we propose a sampling synchronous controller with time-delay ϖ so that the network systems (9.2) can realize the synchronization.

Theorem 9.2 *For a strong digraph $\mathfrak{G}$, given constant delay $\varpi > 0$, upper bound of sampling lag $b_u > 0$ and parameters ϑ_g, $(g = 1, \ldots, 6)$, if exist $\bar{R} > 0$, $\bar{J} > 0$, $\bar{P} > 0$, $\bar{Z}_1 > 0$, $\bar{Z}_2 > 0$, $\bar{Q} > 0$, U, diagonal matrix $\bar{G} > 0$, and diagonal matrix $\bar{H}$, so that*

$$\bar{\mathfrak{M}}^{ij} = \begin{bmatrix} \bar{\mathfrak{M}}_{11} & \bar{\mathfrak{M}}_{12} & \bar{\mathfrak{M}}_{13}^{ij} & \bar{\mathfrak{M}}_{14} & \bar{\mathfrak{M}}_{15} & \bar{\mathfrak{M}}_{16} \\ * & \bar{\mathfrak{M}}_{22} & \bar{\mathfrak{M}}_{23}^{ij} & 0 & 0 & \bar{\mathfrak{M}}_{26} \\ * & * & \bar{\mathfrak{M}}_{33}^{ij} & \bar{\mathfrak{M}}_{34}^{ij} & \bar{\mathfrak{M}}_{35}^{ij} & \bar{\mathfrak{M}}_{36}^{ij} \\ * & * & * & \bar{\mathfrak{M}}_{44} & 0 & \bar{\mathfrak{M}}_{46} \\ * & * & * & * & \bar{\mathfrak{M}}_{55} & \bar{\mathfrak{M}}_{56} \\ * & * & * & * & * & \bar{\mathfrak{M}}_{66} \end{bmatrix} < 0 \tag{9.19}$$

holds for any $i, j = 1, \ldots, N$, where $\bar{\psi}_{ij} = \psi(l_{ij}/\varsigma_j)$ and

$$\begin{aligned}
\bar{\mathfrak{M}}_{11} &= \bar{Q} - \bar{R} + \vartheta_1 A\bar{H} + \vartheta_1 \bar{H}^T A^T + D^T \bar{G} D + \frac{b_u^2}{\varpi \kappa_M} \bar{Z}_2 \\
\bar{\mathfrak{M}}_{12} &= \bar{P} - \vartheta_1 \bar{H} + \vartheta_2 \bar{H}^T A^T \\
\bar{\mathfrak{M}}_{13}^{ij} &= \bar{\psi}_{ij} \vartheta_1 BU + \vartheta_3 \bar{H}^T A^T \\
\bar{\mathfrak{M}}_{14} &= \bar{R} + \vartheta_4 \bar{H}^T A^T - \frac{b_u}{\varpi} \bar{Z}_2 \\
\bar{\mathfrak{M}}_{15} &= \vartheta_5 \bar{H}^T A^T + \frac{b_u}{\kappa_M} \bar{Z}_2 \\
\bar{\mathfrak{M}}_{16} &= \vartheta_1 \bar{H} + \vartheta_6 \bar{H}^T A^T \\
\bar{\mathfrak{M}}_{22} &= b_u^2(\bar{J} + \bar{Z}_2) + \varpi^2 \bar{R} - \vartheta_2 \bar{H} - \vartheta_2 \bar{H}^T \\
\bar{\mathfrak{M}}_{23}^{ij} &= \bar{\psi}_{ij} \vartheta_2 BU - \vartheta_3 \bar{H} \\
\bar{\mathfrak{M}}_{26} &= \vartheta_2 \bar{H} - \vartheta_6 \bar{H}^T \\
\bar{\mathfrak{M}}_{33}^{ij} &= \bar{\psi}_{ij} \vartheta_3 (BU + U^T B^T) - \frac{\pi^2}{4} \bar{J} \\
\bar{\mathfrak{M}}_{34}^{ij} &= \bar{\psi}_{ij} \vartheta_4 U^T B^T + \frac{\pi^2}{4} \bar{J} \\
\bar{\mathfrak{M}}_{35}^{ij} &= \bar{\psi}_{ij} \vartheta_5 U^T B^T \\
\bar{\mathfrak{M}}_{36}^{ij} &= \bar{\psi}_{ij} \vartheta_6 U^T B^T + \vartheta_3 \bar{H} \\
\bar{\mathfrak{M}}_{44} &= \bar{Z}_1 - \bar{Q} - \bar{R} + \frac{b_u}{\varpi} \bar{Z}_2 - \frac{\pi^2}{4} \bar{J} \\
\bar{\mathfrak{M}}_{46} &= \vartheta_4 \bar{H}
\end{aligned}$$

$$\bar{\mathfrak{M}}_{55} = -\bar{Z}_1 - \frac{b_u}{\kappa_M}\bar{Z}_2$$
$$\bar{\mathfrak{M}}_{56} = \vartheta_5\bar{H}$$
$$\bar{\mathfrak{M}}_{66} = \vartheta_6\bar{H} + \vartheta_6\bar{H}^T - \bar{G}$$

the network systems (9.11) *are able to complete synchronization with gain matrix* $K = U\bar{H}^{-1}$ *in controller* (9.8).

Proof Recalling the LMI (9.13), $\mathcal{N}_g = \vartheta_g H$, $g = 1, \ldots, 6$ is chosen, where H is a diagonal matrix. Further, we set

$$\begin{aligned}\bar{H} =& H^{-1},\ U = K\bar{H},\ \bar{P} = \bar{H}^T P\bar{H},\ \bar{Q} = \bar{H}^T Q\bar{H},\ \bar{R} = \bar{H}^T R\bar{H},\\ \bar{J} =& \bar{H}^T J\bar{H},\ \bar{G} = \bar{H}^T G\bar{H},\ \bar{Z}_1 = \bar{H}^T Z_1\bar{H},\ \bar{Z}_2 = \bar{H}^T Z_2\bar{H}.\end{aligned}$$

Pre- and post-multiplying (9.13) by $\mathrm{diag}\{\bar{H}^T, \bar{H}^T, \bar{H}^T, \bar{H}^T, \bar{H}^T, \bar{H}^T\}$ and $\mathrm{diag}\{\bar{H}, \bar{H}, \bar{H}, \bar{H}, \bar{H}, \bar{H}\}$, (9.19) is obtained. □

Remark 9.3 As a practical matter, to get the peak value of the upper bound b_u of sampling intervals usually draws a lot of attention. By addressing the following optimization problem, such issue can be solved:

$$\begin{aligned}&\min\ b_u^{-1}\\ &\text{s. t.}\ \ (9.19)|_{\varpi}\ \text{is given.}\end{aligned} \tag{9.20}$$

By addressing such optimization issue to attain (9.20), the peak value of upper bound b_u is found.

9.3.2 Synchronization via Discretized Lyapunov Functional

In (9.11), if a constant time-delay $\tilde{\varpi} \in [0, \varpi)$ leads to an unstable system (and, therefore, the simple Lyapunov functional V_p is not suitable any more), the Lyapunov functional V_p can be revised as a complete one

$$\begin{aligned}V_{P_1}\big(x(t)\big) =& \ x^T(t)(W\otimes P_1)x(t) + 2x^T(t)\int_{-\varpi}^{0}(W\otimes Q(\gamma))x(t+\gamma)d\gamma\\ &+\int_{-\varpi}^{0}\int_{-\varpi}^{0}x^T(t+\gamma)(W\otimes R(\gamma,\theta))x(\gamma)d\gamma x(t+\theta)d\theta\\ &+\int_{-\varpi}^{0}x^T(t+\gamma)(W\otimes S(\gamma))x(t+\gamma)d\gamma,\quad P_1>0\end{aligned} \tag{9.21}$$

where $Q(s)$, $R(s,\theta)$, and $S(s)$ are piecewise-linear and continuous functions (Gu 1997). According to Gu (1997), the delay interval $[-\varpi, 0]$ can be divided into M segments $[\theta_p, \theta_{p-1}]$, $p = 1, \dots, M$ of the same length $r = \varpi/M$, where $\theta_p = -pr$. Such approach leads to the segmentation of the square $[-\varpi, 0] \times [-\varpi, 0]$ into $M \times M$ small grids $[\theta_p, \theta_{p-1}] \times [\theta_p, \theta_{p-1}]$. Further, every small grid is grouped into a symmetrical triangle pair. We choose the linear matrix functions $Q(s)$, $S(s)$ and $R(s,\theta)$ within each segment or triangle with $p, q = 0, 1, \dots, M$.

$$
\begin{aligned}
&Q(\theta_p + \psi r) = (1-\psi)Q_p + \psi Q_{p-1} \\
&S(\theta_p + \psi r) = (1-\psi)S_p + \psi S_{p-1}, \quad \psi \in [0,1] \\
&R(\theta_p + \psi r, \theta_q + \xi r) = \begin{cases} \xi R_{p-1,q-1} + (1-\psi)R_{pq} + (\psi - \xi)R_{p-1,q}, & \psi \geq \xi \\ \psi R_{p-1,q-1} + (1-\xi)R_{pq} + (\xi - \psi)R_{p,q-1}, & \psi < \xi \end{cases}
\end{aligned}
$$

The Lyapunov functional $V\big(x(t)\big) = V_J\big(x(t)\big) + V_{P_1}\big(x(t)\big) + V_Z\big(x(t)\big)$ is employed. Then, following the arguments of Theorem 9.1 and the descriptor method in Fridman (2006), the following corollary can be achieved.

Corollary 9.1 *For a strong digraph $\mathfrak{G}$, given constant delay $\varpi > 0$, sampling upper bound $b_u > 0$ and gain matrix K in controller, the network nonlinear systems* (9.2) *under Assumption 9.1 are able to complete the asymptotically synchronization, if exist $n \times n$ matrices $P_1 > 0$, $J > 0$, $Z_1 > 0$, $Z_2 > 0$, diagonal matrix $G > 0$, and $\mathcal{N}_s$, $(s = 1, \dots, 6)$, $S_p = S_p^T$, Q_p, $R_{pq} = R_{qp}^T$, $p, q = 1, \dots, M$, so that $\forall l, s = 1, \dots, N$, the following inequalities hold:*

$$
\begin{bmatrix} P_1 & \tilde{Q} \\ * & \tilde{R} + \tilde{S} \end{bmatrix} > 0 \tag{9.22}
$$

$$
\begin{bmatrix} \Xi^{ls} + \Xi_1 & \begin{bmatrix} D^s \\ 0 \end{bmatrix} & \begin{bmatrix} D^a \\ 0 \end{bmatrix} \\ * & -R_d - S_d & 0 \\ * & * & -3S_d \end{bmatrix} < 0 \tag{9.23}
$$

where $r = \varpi/M$, $\bar{\psi}_{ls} = \psi(l_{ls}/\varsigma_s)$ and

$$
\tilde{Q} = \begin{bmatrix} Q_0 \; Q_1 \dots Q_M \end{bmatrix}, \; \tilde{S} = diag\left\{\frac{1}{r}S_0, \frac{1}{r}S_1, \dots, \frac{1}{r}S_M\right\}
$$

$$
\tilde{R} = \begin{bmatrix} R_{00} & R_{01} & \dots & R_{0M} \\ R_{10} & R_{11} & \dots & R_{1M} \\ \vdots & \vdots & \ddots & \vdots \\ R_{M0} & R_{M1} & \dots & R_{MM} \end{bmatrix}
$$

$$\Xi_1 = \begin{bmatrix} Q_0 + Q_0^T + S_0 & 0 & 0 & -Q_N & 0 & 0 \\ * & 0 & 0 & 0 & 0 & 0 \\ * & * & 0 & 0 & 0 & 0 \\ * & * & * & -S_N & 0 & 0 \\ * & * & * & * & 0 & 0 \\ * & * & * & * & * & 0 \end{bmatrix}$$

$$\Xi^{ls} = \begin{bmatrix} \Xi_{11} & \Xi_{12} & \Xi_{13}^{ls} & \Xi_{14} & \Xi_{15} & \Xi_{16} \\ * & \Xi_{22} & \Xi_{23}^{ls} & 0 & 0 & \Xi_{26} \\ * & * & \Xi_{33}^{ls} & \Xi_{34}^{ls} & \Xi_{35}^{ls} & \Xi_{36}^{ls} \\ * & * & * & \Xi_{44} & 0 & \Xi_{46} \\ * & * & * & * & \Xi_{55} & \Xi_{56} \\ * & * & * & * & * & \Xi_{66} \end{bmatrix}$$

$$\Xi_{11} = \mathcal{N}_1 A + A^T \mathcal{N}_1^T + D^T G D + \frac{b_u^2}{\varpi \kappa_M} Z_2$$

$$\Xi_{12} = P_1 - \mathcal{N}_1 + A^T \mathcal{N}_2^T, \ \Xi_{13}^{ls} = \bar{\psi}_{ls} \mathcal{N}_1 B K + A^T \mathcal{N}_3^T$$

$$\Xi_{14} = A^T \mathcal{N}_4^T - \frac{b_u}{\varpi} Z_2, \ \Xi_{15} = A^T \mathcal{N}_5^T + \frac{b_u}{\kappa_M} Z_2$$

$$\Xi_{16} = \mathcal{N}_1 + A^T \mathcal{N}_6^T, \ \Xi_{22} = b_u^2 (J + Z_2) - \mathcal{N}_2 - \mathcal{N}_2^T$$

$$\Xi_{23}^{ls} = \bar{\psi}_{ls} \mathcal{N}_2 B K - \mathcal{N}_3, \ \Xi_{26} = \mathcal{N}_2 - \mathcal{N}_6^T$$

$$\Xi_{33}^{ls} = \bar{\psi}_{ls} (\mathcal{N}_3 B K + K^T B^T \mathcal{N}_3^T) - \frac{\pi^2}{4} J$$

$$\Xi_{34}^{ls} = \bar{\psi}_{ls} K^T B^T \mathcal{N}_4^T + \frac{\pi^2}{4} J, \ \Xi_{35}^{ls} = \bar{\psi}_{ls} K^T B^T \mathcal{N}_5^T$$

$$\Xi_{36}^{ls} = \bar{\psi}_{ls} K^T B^T \mathcal{N}_6^T + \mathcal{N}_3, \ \Xi_{44} = Z_1 + \frac{b_u}{\varpi} Z_2 - \frac{\pi^2}{4} J$$

$$\Xi_{46} = \mathcal{N}_4, \ \Xi_{55} = -Z_1 - \frac{b_u}{\kappa_M} Z_2$$

$$\Xi_{56} = \mathcal{N}_5, \ \Xi_{66} = \mathcal{N}_6 + \mathcal{N}_6^T - G$$

$$\phi^T(\psi) = \left[x^T(t - h + \psi h) \, x^T(t - 2r + \psi r) \dots x^T(t - Mr + \psi r) \right]$$

$$S_d = diag\{S_0 - S_1, S_1 - S_2, \dots, S_{M-1} - S_M\}$$

$$R_d = \begin{bmatrix} R_{d11} & R_{d12} & \cdots & R_{d1M} \\ R_{d21} & R_{d22} & \cdots & R_{d2M} \\ \vdots & \vdots & \ddots & \vdots \\ R_{dM1} & R_{dM2} & \cdots & R_{dMM} \end{bmatrix}, \ D_p^a = \begin{bmatrix} -r/2(R_{0,p-1} - R_{0p}) \\ -r/2(Q_{p-1} - Q_p) \\ r/2(R_{M,p-1} - R_{M_p}) \end{bmatrix}$$

$$D_p^s = \begin{bmatrix} r/2(R_{0,p-1} + R_{0p}) - (Q_{p-1} - Q_p) \\ r/2(Q_{p-1} + Q_p) \\ -r/2(R_{M,p-1} + R_{M_p}) \end{bmatrix}$$

$$D^a = \left[D_1^a \ D_2^a \dots D_M^a \right], \ D^s = \left[D_1^s \ D_2^s \dots D_M^s \right]$$

$$R_{dpq} = r(R_{p-1,q-1} - R_{pq}).$$

Proof Differentiating Lyapunov functional (9.21) gives rise to

$$\begin{aligned}\dot{V}_{P_1}\big(x(t)\big) =\ & 2\dot{x}^T(t)\Big[(W\otimes P_1)x(t)+\int_{-\varpi}^{0}(W\otimes Q(\gamma))x(t+\gamma)d\gamma\Big]\\ &+2x^T(t)\int_{-\varpi}^{0}(W\otimes Q(\gamma))\dot{x}(t+\gamma)d\gamma\\ &+2\int_{-\varpi}^{0}\int_{-\varpi}^{0}\dot{x}^T(t+\gamma)(W\otimes R(\gamma,\theta))d\gamma x(t+\theta)d\theta\\ &+2\int_{-\varpi}^{0}\dot{x}^T(t+\gamma)(W\otimes S(\gamma))x(t+\gamma)d\gamma. \end{aligned}\tag{9.24}$$

In order to remove Kronecker product, utilizing Lemma 9.2 onto (9.24) results in

$$\begin{aligned}\dot{V}_{P_1}\big(x(t)\big) =\ & \frac{1}{2}\sum_{l=1}^{N}\sum_{s=1,s\neq l}^{N}\varsigma_l\varsigma_s\Big\{2\dot{\widetilde{x}}^T(t)P_1\widetilde{x}(t)\\ &+2\widetilde{x}^T(t)\int_{-\varpi}^{0}Q(\gamma)\dot{\widetilde{x}}(t+\gamma)d\gamma+2\dot{\widetilde{x}}^T(t)\int_{-\varpi}^{0}Q(\gamma)\widetilde{x}(t+\gamma)d\gamma\\ &+2\int_{-\varpi}^{0}\int_{-\varpi}^{0}\dot{\widetilde{x}}^T(t+\gamma)R(\gamma,\theta)\widetilde{x}(t+\theta)d\gamma d\theta\\ &+2\int_{-\varpi}^{0}\dot{\widetilde{x}}^T(t+\gamma)S(\gamma)\widetilde{x}(t+\gamma)d\gamma\Big\}. \end{aligned}\tag{9.25}$$

Considering the Lyapunov functional $V = V_J + V_{P_1} + V_Z$, from (9.25), $\dot{V}_J$, $\dot{V}_Z$, (9.15) and (9.16), we obtain

$$\begin{aligned}\dot{V}\big(x(t)\big) \le\ & \frac{1}{2}\sum_{l=1}^{N}\sum_{s=1,s\neq l}^{N}\varsigma_l\varsigma_s\Big\{\widetilde{\xi}^T(t)\Xi^{ls}\widetilde{\xi}(t)\\ &+2\widetilde{x}^T(t)\int_{-\varpi}^{0}Q(\gamma)\dot{\widetilde{x}}(t+\gamma)d\gamma+2\dot{\widetilde{x}}^T(t)\int_{-\varpi}^{0}Q(\gamma)\widetilde{x}(t+\gamma)d\gamma \end{aligned}$$

$$+2\int_{-\varpi}^{0}\int_{-\varpi}^{0}\dot{\tilde{x}}^{T}(t+\gamma)R(\gamma,\theta)d\gamma\tilde{x}(t+\theta)d\theta$$

$$+2\int_{-\varpi}^{0}\dot{\tilde{x}}^{T}(t+\gamma)S(\gamma)\tilde{x}(t+\gamma)d\gamma\Big\}. \tag{9.26}$$

Integrating by parts in (9.26), we have

$$\dot{V}(x(t)) \leq \frac{1}{2}\sum_{l=1}^{N}\sum_{s=1,s\neq l}^{N}\varsigma_l\varsigma_s\Big\{\widetilde{\xi}^{T}(t)\big(\Xi^{ls}+\Xi_0\big)\widetilde{\xi}(t)$$

$$+2\widetilde{\dot{x}}^{T}(t)\int_{-\varpi}^{0}Q(\gamma)\widetilde{x}(t+\gamma)d\gamma$$

$$-\int_{-\varpi}^{0}\int_{-\varpi}^{0}\widetilde{x}^{T}(t+\gamma)\Big[\Big(\frac{\partial}{\partial\gamma}+\frac{\partial}{\partial\theta}\Big)R(\gamma,\theta)\Big]\widetilde{x}(t+\theta)d\gamma d\theta$$

$$+2\widetilde{x}^{T}(t)\int_{-\varpi}^{0}\Big[-\dot{Q}(\gamma)+R(0,\gamma)\Big]\widetilde{x}(t+\gamma)d\gamma$$

$$-2\widetilde{x}^{T}(t-\varpi)\int_{-\varpi}^{0}R(-\varpi,\theta)\widetilde{x}(t+\theta)d\theta$$

$$-\int_{-\varpi}^{0}\widetilde{x}^{T}(t+\gamma)\dot{S}(\gamma)\widetilde{x}(t+\gamma)d\gamma\Big\} \tag{9.27}$$

where

$$\Xi_0=\begin{bmatrix} Q(0)+Q^T(0)+S(0) & 0 & 0 & -Q(-\varpi) & 0 & 0\\ * & 0 & 0 & 0 & 0 & 0\\ * & * & 0 & 0 & 0 & 0\\ * & * & * & -S(-\varpi) & 0 & 0\\ * & * & * & * & 0 & 0\\ * & * & * & * & * & 0\end{bmatrix}.$$

Let $r=\varpi/M$. The Lyapunov-Krasovskii functional (LKF) condition $V(x(t))\geq \varepsilon\|x(t)\|^2$, $\varepsilon>0$ can be met if $S_p>0$, $p=0,1,\ldots,M$ and

$$\begin{bmatrix} P_1 & \tilde{Q}\\ * & \tilde{R}+\tilde{S}\end{bmatrix}>0. \tag{9.28}$$

To impel the derivation of LKF derivative condition, it can be noted that

$$\begin{aligned}\dot{S}(\gamma) &= \frac{1}{h}(S_{p-1} - S_p) \\ \dot{Q}(\gamma) &= \frac{1}{h}(Q_{p-1} - Q_p) \\ \frac{\partial}{\partial \gamma}R(\gamma, \theta) + \frac{\partial}{\partial \theta}R(\gamma, \theta) &= \frac{1}{h}(R_{p-1,q-1} - R_{pq}). \end{aligned} \tag{9.29}$$

Therefore

$$\begin{aligned}&2\tilde{\dot{x}}^T(t)\int_{-\varpi}^{0} Q(\gamma)\widetilde{x}(t+\gamma)d\gamma \\ &= 2\widetilde{\dot{x}}^T(t)\sum_{p=1}^{N} r\int_{0}^{1}[(1-\psi)Q_p + \psi Q_{p-1}]\widetilde{x}(t+\theta_p+\psi r)d\psi \\ &= 2\widetilde{\dot{x}}^T(t)\sum_{p=1}^{N} r\int_{0}^{1}[(1-\psi)(Q_p^s + Q_p^a) + \psi(Q_p^s - Q_p^a)]\widetilde{x}(t+\theta_p+\psi r)d\psi\end{aligned}$$

where $Q_p^s = 1/2(Q_{p-1} + Q_p)$, $Q_p^a = 1/2(Q_p - Q_{p-1})$.

Equations (9.27) and (9.29) imply that

$$\begin{aligned}\dot{V}(x(t)) \leq \frac{1}{2}\sum_{l=1}^{N}\sum_{s=1,s\neq l}^{N} \varsigma_l\varsigma_s\Big\{&\widetilde{\xi}^T(t)\big(\Xi^{ls} + \Xi_1\big)\widetilde{\xi}(t) \\ &- \int_0^1 \phi^T(\psi)S_d\phi(\psi)d\psi - \int_0^1\Big[\int_0^1 \phi^T(\psi)R_d\phi(\xi)d\psi\Big]d\xi \\ &+ 2\widetilde{\xi}^T(t)\int_0^1\Big[D^s + (1-2\psi)D^a\Big]\phi(\psi)d\psi\Big\}. \end{aligned} \tag{9.30}$$

Utilizing Prop. 5.21 in Gu et al. (2003) onto (9.30), we can summarize that $\dot{V}(x(t)) \leq 0$ if (9.23) holds.

9.4 Numerical Example

This section provides a simulation together with practical example to testify the efficacy of the designed approach.

The network UGV systems can be practically used for network system theory. Our algorithms have been utilized to control the UGVs in our laboratory, shown in Fig. 9.1, and some satisfactory results have been achieved.

In Fig. 9.2, the topological structure $\mathfrak{G}$ of UGVs is shown. Then, the corresponding adjacent matrix $\mathfrak{A}$ and the Laplacian matrix $\mathfrak{L}$ are provided as

$$\mathfrak{A} = \begin{bmatrix} 0\,1\,0\,0\,1 \\ 1\,0\,1\,0\,0 \\ 0\,1\,0\,1\,0 \\ 0\,0\,1\,0\,1 \\ 1\,0\,0\,1\,0 \end{bmatrix}, \quad \mathfrak{L} = \begin{bmatrix} 2 & -1 & 0 & 0 & -1 \\ -1 & 2 & -1 & 0 & 0 \\ 0 & -1 & 2 & -1 & 0 \\ 0 & 0 & -1 & 2 & -1 \\ -1 & 0 & 0 & -1 & 2 \end{bmatrix}.$$

We can see that $\mathfrak{G}$ is a strong digraph with the normalized left eigenvector $\varsigma = \left[1/5\ 1/5\ 1/5\ 1/5\ 1/5\right]^T$.

Ultra-Wideband (UWB) positioning system is utilized to support data communication of five UGVs. UWB positioning tag modules are installed on UGVs to obtain their positioning information. Moreover, the ZigBee module of a UGV can send its own information and receive information from other UGVs. Four base transceiver station (BTS) towers A_0, A_1, A_2 and A_3, which are shown in Fig. 9.3, are used to build a communication network. Each BTS tower is placed in each corner of a rectangular area. The UWB positioning system consists of BTS towers and positioning tag modules. UGVs with positioning tag modules can get their own position coordinates

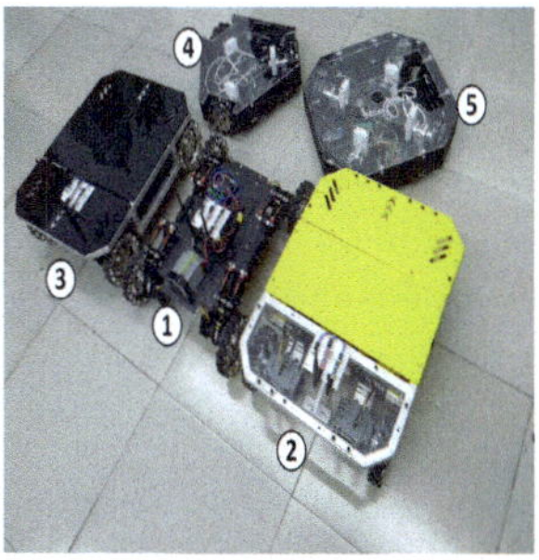

Fig. 9.1 The network systems with five UGVs

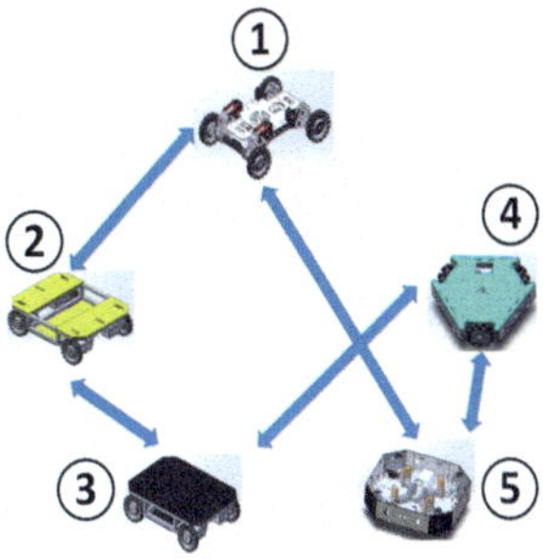

Fig. 9.2 Topological structure of the network UGV systems

Fig. 9.3 Four BTS towers labeled by A_0, A_1, A_2 and A_3

in the area surrounded by the four BTS towers. Indoor UWB positioning system can obtain the location information of UGVs with high accuracy, which is less than 10 cm.

We can describe the dynamic characteristics of the ith UGV by Eq. (9.2) with the following parameter:

$$A = \begin{bmatrix} 0 & -0.9973 \\ 0.1613 & 0 \end{bmatrix}$$

$$B = \begin{bmatrix} -6.5632 & -6.1019 \\ -6.9936 & 0.1084 \end{bmatrix}$$

$$f(x_l(t)) = \begin{bmatrix} 0.5\cos(x_{l1}(t)) \; 0.5\sin(x_{l2}(t)) \end{bmatrix}^T$$

where $x_l(t) = \begin{bmatrix} x_{l1}(t) \; x_{l2}(t) \end{bmatrix}^T$, $l = 1, \ldots, 5$. Reviewing Assumption 9.1, it can be seen that the nonlinear property of each UGV satisfies the global Lipschitz condition with $d_1 = d_2 = 0.5$.

The initial conditions of the five UGVs are given as follows, which represent the position coordinates of each UGV before starting the experiment.

$$x_1(0) = \begin{bmatrix} -15 \\ 3 \end{bmatrix}, \quad x_2(0) = \begin{bmatrix} 8 \\ 15 \end{bmatrix}, \quad x_3(0) = \begin{bmatrix} -5 \\ 8 \end{bmatrix}$$

$$x_4(0) = \begin{bmatrix} 15 \\ -8 \end{bmatrix}, \quad x_5(0) = \begin{bmatrix} -8 \\ -15 \end{bmatrix}.$$

It can be seen that the five UGVs are placed dispersedly before experiment in Fig. 9.4.

Recalling the framework of the sampled data controller (9.8), it is supposed that the coupling strength is $\psi = 0.5$. The constant delay is $\varpi = 0.2$, sampling upper bound is $b_u = 0.8$. Hence, we have $\kappa_M = b_u + \varpi = 1$. According to Theorem 9.2, we can obtain the gain matrix K of controller as

$$K = \begin{bmatrix} -0.0112 & -0.0473 \\ -0.0519 & 0.1143 \end{bmatrix}.$$

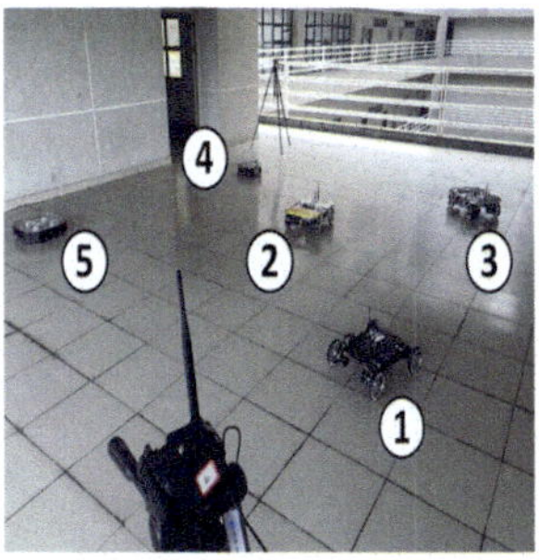

Fig. 9.4 The five UGVs are placed dispersedly before experiment

Putting K and ϖ into the sampled data controller (9.8), the state trajectories of the five UGVs with such sampling synchronous controllers are shown in Fig. 9.5.

It can be observed that the five UGVs come together and move to the same coordinate point in Fig. 9.6. In order to avoid collision among UGVs, slight differences are inevitable in their coordinates.

Figure 9.7 gives the distribution of sampling intervals. The y-axis value of each stem is $(t_{k+1} - t_k)/b_u$, and the x-axis value is the sampling instant t_k. It can be found from Fig. 9.7 that the number of communications decreased significantly than continuous situation. Additionally, a larger sampling upper bound $b_u = 0.99$ is obtained, which is larger than 0.68 in Wu et al. (2016).

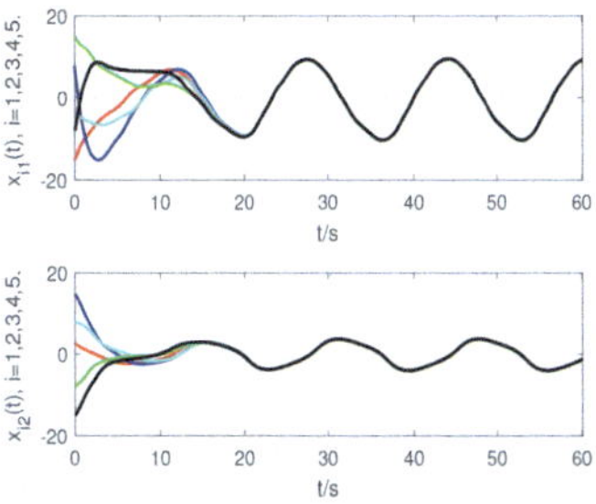

Fig. 9.5 State trajectories of the five UGVs

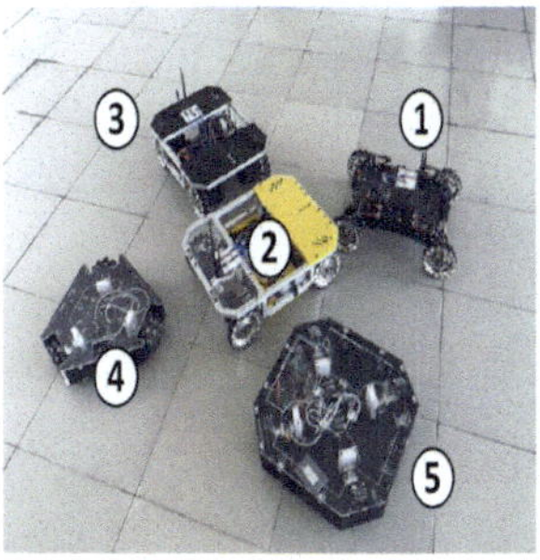

Fig. 9.6 The network UGV systems achieve synchronization

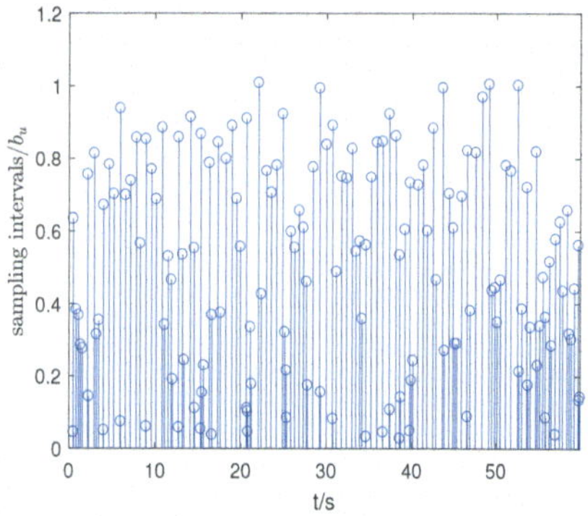

Fig. 9.7 Aperiodic sampling intervals with upper bound b_u

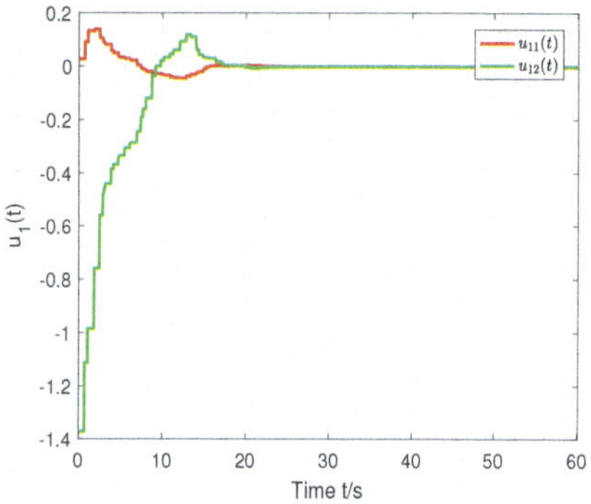

Fig. 9.8 Sampled data control signal $u_1(t)$ for UGV 1

In Fig. 9.8, the sampling control signals with ZOH for UGV 1 are given. From (9.8), it can be found that $u_i(t), t \in [t_k, t_{k+1})$ use the information of the states $x_i(t_k - \varpi)$ with constant delay ϖ. Furthermore, the signals $u_i(t), t \in [t_k, t_{k+1})$ remain the same during the sampling intervals $[t_k, t_{k+1})$ due to the ZOH.

In order to further demonstrate the effectiveness of our proposed algorithm, for every UGV, the synchro error is given by $\epsilon(t) = \sqrt{\sum_{j=2}^{5} \|x_j(t) - x_1(t)\|}$. The synchronization error $\epsilon(t)$ is presented in Fig. 9.9 with a fast convergence rate which can be seen that the synchronization of five UGVs is achieved within a short time.

For each UGV, the trajectories of the norm $\|x_i(t)\|$ are presented in Fig. 9.10. It can be seen that the five UGVs achieve the synchronization by the designed controller.

From Fig. 9.11, we can see that five UGVs cannot achieve synchronization with out control input.

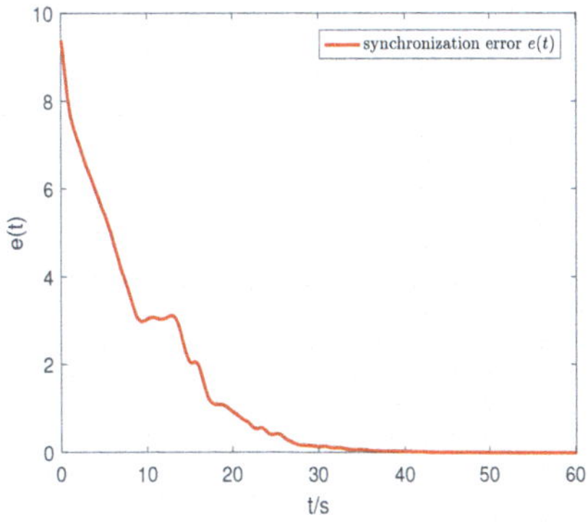

Fig. 9.9 Synchronization error $\epsilon(t)$ attenuate to 0 exponentially

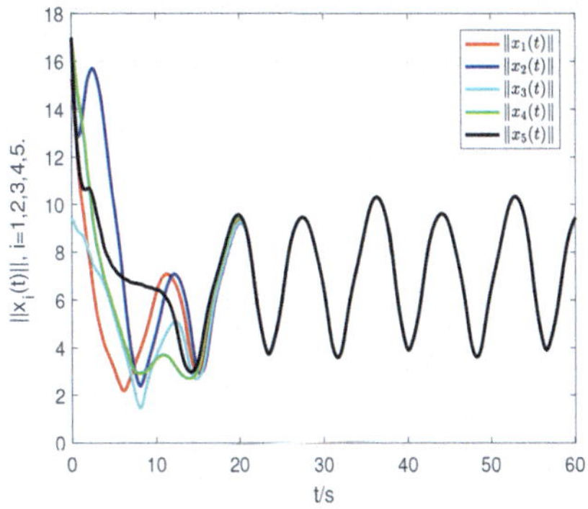

Fig. 9.10 Norm of the state trajectories $\|x_i(t)\|$ of the five UGVs

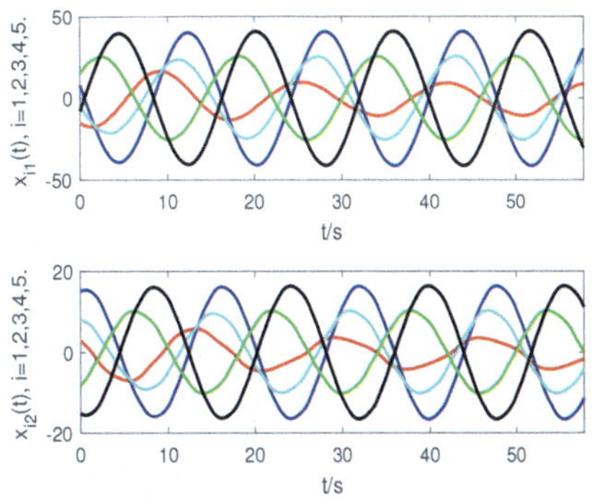

Fig. 9.11 State trajectories of the five UGVs without control input

9.5 Conclusions

In the chapter, synchronization issues have been investigated for network nonlinear systems with constant delay. Together with ZOH, the aperiodic sampled data control law has been applied. Compared with the traditional periodic sampled data control method, this approach has demonstrated more greater flexibility. Adopting input delay method, we have remodeled the sampling system as continuous system with time-varying delays. For the case of the sampled data controller suffering a constant input delay, an updated Lyapunov functional has been applied derived from the augmented Wirtinger's inequality for stable network systems. This Lyapunov functional has resulted in efficient and simplified stability conditions. A sufficient condition for synchronizability of network systems has been established. Then, for the situation of unstable network systems with some constant delay, a novel discretized Lyapunov functional has been proposed. Finally, an example of numerical simulation has been provided to testify the efficacy and advantage of the designed method. Moreover, the proposed algorithm has been verified in a real scenario based on the network UGV systems. In future work, autonomous formation control of UGVs will be studied.

References

Ali MS, Vadivel R (2018) Decentralized event-triggered exponential stability for uncertain delayed genetic regulatory networks with Markov jump parameters and distributed delays. Neural Process Lett 47(3):1219–1252

Ali MS, Vadivel R, Kwon O (2018a) Decentralized event-triggered stability analysis of neutral-type BAM neural networks with Markovian jump parameters and mixed time varying delays. Int J Control Autom Syst 16(3):983–993

Ali MS, Gunasekaran N, Kwon O (2018b) Delay-dependent H_∞ performance state estimation of static delayed neural networks using sampled-data control. Neural Comput Appl 30(2):539–550

Chen CLP, Wen GX, Liu YJ, Wang FY (2014) Adaptive consensus control for a class of nonlinear multiagent time-delay systems using neural networks. IEEE Trans Neural Netw Learn Syst 25(6):1217–1226

Deng F, Guan S, Yue X, Gu X, Chen J, Lv J, Li J (2017a) Energy-based sound source localization with low power consumption in wireless sensor networks. IEEE Trans Indus Electron 64(6):4894–4902

Deng F, Guo S, Zhou R, Chen J (2017b) Sensor multifault diagnosis with improved support vector machines. IEEE Trans Autom Sci Eng 14(2):1053–1063

Deng F, Ma L, Gao X, Chen J (2019) The MR-CA models for analysis of pollution sources and prediction of PM2.5. IEEE Trans Syst Man Cybernet Syst 49(4):814–820

Fridman E (2006) Descriptor discretized Lyapunov functional method: analysis and design. IEEE Trans Autom Control 51(5):890–897

Fridman E (2010) A refined input delay approach to sampled-data control. Automatica 46(2):421–427

Gao H, Wu J, Shi P (2009) Robust sampled-data H_∞ control with stochastic sampling. Automatica 45(7):1729–1736

Gu K (1997) Discretized LMI set in the stability problem of linear uncertain time-delay systems. Int J Control 68(4):923–934

Gu K, Chen J, Kharitonov V (2003) Stability of time-delay systems. Birkhauser Boston

Guo M, Dimarogonas DV (2013) Nonlinear consensus via continuous, sampled, and aperiodic updates. Int J Control 86(4):567–578

Karimi HR, Duffie NA, Dashkovskiy S (2010) Local capacity H_∞ control for production networks of autonomous work systems with time-varying delays. IEEE Trans Autom Sci Eng 7(4):849–857

Li Z, Duan Z, Chen G (2011) Dynamic consensus of linear multi-agent systems. IET Control Theory Appl 5(1):19–28

Li H, Jing X, Karimi HR (2013) Output-feedback-based H_∞ control for vehicle suspension systems with control delay. IEEE Trans Indus Electron 61(1):436–446

Liang H, Zhang Y, Huang T, Ma H (2020) Prescribed performance cooperative control for multiagent systems with input quantization. IEEE Trans Cybernet 50(5):1810–1819
Lin P, Jia Y (2010) Consensus of a class of second-order multi-agent systems with time-delay and jointly-connected topologies. IEEE Trans Autom Control 55(3):778–784
Lin P, Jia Y, Li L (2008) Distributed robust H_∞ consensus control in directed networks of agents with time-delay. Syst Control Lett 57(8):643–653
Liu K, Fridman E (2012) Wirtinger's inequality and Lyapunov-based sampled-data stabilization. Automatica 48(1):102–108
Liu K, Fridman E (2014) Delay-dependent methods and the first delay interval. Syst Control Lett 64(1):57–63
Liu ZW, Guan ZH, Shen X, Feng G (2012) Consensus of multi-agent networks with aperiodic sampled communication via impulsive algorithms using position-only measurements. IEEE Trans Autom Control 57(10):2639–2643
Liu K, Fridman E, Hetel L (2012) Stability and L_2-gain analysis of networked control systems under round-robin scheduling: a time-delay approach. Syst Control Lett 61(5):666–675
Liu S, Li T, Xie L, Fu M, Zhang JF (2013) Continuous-time and sampled-data-based average consensus with logarithmic quantizers. Automatica 49(11):3329–3336
Lu J, Ho DW (2010) Globally exponential synchronization and synchronizability for general dynamical networks. IEEE Trans Syst Man Cybernet Part B: Cybernet 40(2):350–361
Lu R, Shi P, Su H, Wu ZG, Lu J (2016) Synchronization of general chaotic neural networks with nonuniform sampling and packet missing: a switched system approach. IEEE Trans Neural Netw Learn Syst 29(3):523–533
Ma Q, Lu J, Xu H (2014) Consensus for nonlinear multi-agent systems with sampled data. Trans Inst Meas Control 36(5):618–626
Ma Z, Liu Z, Chen Z (2017) Modified leader-following consensus of time-delay multi-agent systems via sampled control and smart leader. Int J Control Autom Syst 15(6):2526–2537
Ma H, Li H, Liang H, Dong G (2019) Adaptive fuzzy event-triggered control for stochastic nonlinear systems with full state constraints and actuator faults. IEEE Trans Fuzzy Syst 27(11):2242–2254
Meng X, Chen T (2012) Optimal sampling and performance comparison of periodic and event based impulse control. IEEE Trans Autom Control 57(12):3252–3259
Mu B, Shi Y (2018) Distributed LQR consensus control for heterogeneous multi-agent systems: theory and experiments. IEEE/ASME Trans Mechatron 23(1):434–443
Oishi Y, Fujioka H (2010) Stability and stabilization of aperiodic sampled-data control systems using robust linear matrix inequalities. Automatica 46(8):1327–1333
Ren W, Beard RW (2005) Consensus seeking in multiagent systems under dynamically changing interaction topologies. IEEE Trans Autom Control 50(5):655–661
Semsar-Kazerooni E, Khorasani K (2010) Optimal consensus seeking in a network of multiagent systems: an LMI approach. IEEE Trans Syst Man Cybernet Part B: Cybernet 40(2):540–547
Shen C, Shi Y, Buckham B (2018) Trajectory tracking control of an autonomous underwater vehicle using Lyapunov-based model predictive control. IEEE Trans Indus Electron 65(7):5796–5805
Shi P (1998) Filtering on sampled-data systems with parametric uncertainty. IEEE Trans Autom Control 43(7):1022–1027
Shi Y, Yu B (2009) Output feedback stabilization of networked control systems with random delays modeled by markov chains. IEEE Trans Autom Control 54(7):1668–1674
Shi Y, Yu B (2011) Robust mixed H_2/H_∞ control of networked control systems with random time delays in both forward and backward communication links. Automatica 47(4):754–760
Suh YS (2008) Stability and stabilization of nonuniform sampling systems. Automatica 44(12):3222–3226
Sun Q, Han R, Zhang H, Zhou J, Guerrero JM (2015) A multiagent-based consensus algorithm for distributed coordinated control of distributed generators in the energy internet. IEEE Trans Smart Grid 6(6):3006–3019

Sun W, Su SF, Wu Y, Xia J, Nguyen VT (2020) Adaptive fuzzy control with high-order barrier Lyapunov functions for high-order uncertain nonlinear systems with full-state constraints. IEEE Trans Cybernet 50(8):3424–3432
Thunberg J, Song W, Montijano E, Hong Y, Hu X (2014) Distributed attitude synchronization control of multi-agent systems with switching topologies. Automatica 50(3):832–840
Trentelman HL, Takaba K, Monshizadeh N (2013) Robust synchronization of uncertain linear multi-agent systems. IEEE Trans Autom Control 58(6):1511–1523
Wu Y, Su H, Shi P, Shu Z, Wu ZG (2016) Consensus of multiagent systems using aperiodic sampled-data control. IEEE Trans Cybernet 46(9):2132–2143
Wu Y, Karimi HR, Lu R (2018) Sampled-data control of network systems in industrial manufacture. IEEE Trans Syst Man Cybernet Syst 65(11):9016–9024
Xia J, Zhang J, Feng J, Wang Z, Zhuang G (2021) Command filter-based adaptive fuzzy control for nonlinear systems with unknown control directions. IEEE Trans Syst Man Cybernet Syst 51(3):1945–1953
Xiong W, Ho DWC, Cao J (2012) Impulsive consensus of multi-agent directed networks with nonlinear perturbations. Int J Robust Nonlinear Control 22(14):1571–1582
Yan Y, Huang J (2017) Cooperative output regulation of discrete-time linear time-delay multi-agent systems under switching network. Neurocomputing 241:108–114
Zhang Y, Tian YP (2010) Consensus of data-sampled multi-agent systems with random communication delay and packet loss. IEEE Trans Autom Control 55(4):939–943
Zhang H, Shi Y, Wang J, Chen H (2018) A new delay-compensation scheme for networked control systems in controller area networks. IEEE Trans Indus Electron 65(9):7239–7247
Zhang Z, Liang H, Wu C, Ahn CK (2019) Adaptive event-triggered output feedback fuzzy control for nonlinear networked systems with packet dropouts and actuator failure. IEEE Trans Fuzzy Syst 27(9):1793–1806

The manufacturer's authorised representative in the EU is Springer Nature Customer Service Centre GmbH, Europaplatz 3, 69115 Heidelberg, Germany. If you have any concerns regarding our products, please contact ProductSafety@springernature.com

Printed and bound by CPI Group (UK) Ltd, Croydon, CR0 4YY
01/10/2026
02225945-0001